FACTORS OF SOIL FORMATION

FACTORS OF SOIL FORMATION

A System of Quantitative Pedology

HANS JENNY

FOREWORD BY
RONALD AMUNDSON
University of California, Berkeley

DOVER PUBLICATIONS, INC.
New York

Copyright

Bibliographical Note

This Dover Edition, first published in 1994, is an unabridged, unaltered republication of the work first published by the McGraw-Hill Book Company, Inc., 1941. Ronald Amundson, Associate Professor of Soil Science at the University of California, Berkeley, has written a Foreword specially for this edition.

Library of Congress Cataloging-in-Publication Data

Jenny, Hans, 1899–1992.
Factors of soil formation : a system of quantitative pedology / Hans Jenny ; foreword by Ronald Amundson.
p. cm.
Originally published: New York : McGraw-Hill, 1941. With new foreword.
Includes bibliographical references and index.
ISBN-13: 978-0-486-68128-3
ISBN-10: 0-486-68128-9
1. Soil formation. I. Title.
S592.2.J455 1994
551.3'05—dc20 94-17606
CIP

Manufactured in the United States by LSC Communications
68128908 2022
www.doverpublications.com

To

MY WIFE

FOREWORD TO THE DOVER EDITION

Giants were on the earth in those days
—*Genesis 6:4*

Hans Jenny was born 1899 in Basel, Switzerland, and died 92 years later near his adopted home of Berkeley, California. Between these two moments in time lies a life lived deliberately. It would be difficult to state when Jenny's career in science began, but it is true to say it ended only with his death. His *oeuvre* ranks among those of the giants in the earth sciences. Even within this rarefied atmosphere of intellectual achievements, his book *Factors of Soil Formation* stands out as a masterpiece.

Factors of Soil Formation, subtitled *A System of Quantitative Pedology*, is not a textbook, as is commonly stated, but is instead a scientific methodology. More fundamentally, it is a way of seeing the physical world. Within any science there are few books of this type, and within a science as young as pedology, there are arguably no others of such stature. Like Darwin's *Origin of Species* or Lyell's *Principles of Geology*, Jenny's *Factor's of Soil Formation* is, to use Darwin's description (1), "one long argument."

What is the "long argument" contained in *Factors of Soil Formation?* The book is a detailed discussion of the nature of the earth's terrestrial environment and a method of subdividing and studying it. First, soils are recognized as being part of a continuum of components at the earth's surface. Therefore, the distinction between soil and environment is an arbitrary, human construct. Second, given this statement, it is nonetheless possible to subdivide this continuum into systems, or units of study, whose boundaries can be quantitatively, although somewhat arbitrarily, defined—both in time and space. Third, the properties of the system in question "may be described accurately by indicating the initial state of the system, the reaction time, and conditioning variables" (2, p. 14), which Jenny combined and designated as "independent variables" or "soil-forming" factors. These factors are independent of the system being studied, and can be made to vary independently of one another through the judicious selection of study sites. Chapter I consists of a detailed and rigorous discussion of these points. It is impossible to summarize briefly the contents of this critical chapter, and an appreciation of its subtleties and complexities is the reward of repeated readings. One of the greatest simplifications, and misrepresentations, of this chapter and the book in general has been to present Jenny's complex theory as simply:

$$s = f(cl, o, r, p, t, \ldots),$$

which is equation 4 in *Factors*. As Jenny lamented many times in later years, "the equation looks simple, but it's not." Indeed, the bulk of the book, Chapters III through VIII, is devoted to a deeper theoretical definition of each state factor, and empirical data selected to illustrate the principles of the theory. Chapter II, "Methods of Presentation of Soil Data," is a seemingly curious inclusion that summarizes various ways of numerically representing soil properties. While the argument of this chapter might seem self-evident to us today, it was critical to Jenny's vision that soil properties be measured quantitatively, and expressed numerically, thus allowing them to be mathematically correlated to the various state factors.

Jenny's book presents a factorial, or "formalistic" (2, p. 18), theory whose functional relationships are empirically derived from either experimental or observational measurements of soils. As a result, it does not tell us about mechanisms, or types, of pedogenic processes; "it is phenomenological" (3). In order to answer questions regarding mechanisms, one must resort to different theories. The state-factor theory *is*, however, quantitative. Numerous graphical relationships, many of them mathematically expressed, dominate the book. As such, *Factors of Soil Formation* provides an impressive compilation of present-day soil property/state-factor relationships and provides the quantitative data needed for a "uniformitarian" analysis of soils of the past (4) or for predicting the role of soils in ameliorating environmental effects of human activity.

We are commonly introduced to important scientists through textbooks. In photographs, invariably taken during the latter stages of the person's career, the scientist stares rigidly at us from the page. Their careers, and lives, summarized in a few paragraphs, portray a life of duty, perseverance and sacrifice. None of this, of course, gives us a sense of the human spirit that guided the intellect.

It was my good fortune that my life intersected with that of Hans Jenny when I was still a young man. While a graduate student, on a trip that brought me near Berkeley, I spontaneously made an appointment to meet him in his office on the Berkeley campus. That rainy day, in the winter of 1982, was the first of what were to be many visits I made to his small, cramped retirement quarters in 118 Giannini Hall. Sitting on the wicker lawn chair, reserved for his office guests, I was questioned, in the best Socratic tradition, about the conceptual aspects of my doctoral research. Topics of discussion were never selected trivially with Hans, for it was one of his most frequent comments about pedology (consistently he seems to have applied it to his own life) that "we need to discuss theories more."

And so we did. And, probably, so did anyone who wandered into his office, regardless of their original intent. In his gentle way, Hans would guide himself, and his visitor, through a rigorous analysis of the logic of his or her beliefs. The Hans Jenny I knew was a man whose intellect was matched by humility, seasoned with a dry wit and sense of humor. Thoroughly happy, standing before a large audience honoring him at his eighty-fifth birthday, he said, "I disagree with much that has been said about me, for all that I think I can say about myself is that I tried to understand nature." Then, with a large smile on his face, he added, "I thank you for what you have said about me—true or untrue. I feel that it was much nicer to hear these sentiments expressed at an occasion such as this than in an obituary column. I'll remember this all one day, when I'm out there pushing up daisies."

I found it difficult to spur him into providing a personal background to his work. Tidbits, offered rarely, were captivating. Over a period of several years interviews by two individuals became part of his oral history (5), a text of over three hundred pages. Although clearly pleased with its publication in 1989, he remarked to me, "I don't think it's very good," primarily from, it now seems to me, a self-deprecating response to all the attention. Yet, in true Jenny style, he made a lively speech at the dedication ceremony for the history, humorously remarking that "as I get older and live long enough, most of my former enemies are gone and I'm getting all sorts of recognition and awards for my contributions."

It has been stated that a great scientific achievement is the rare result of the right combination of person, place and time (6). It is intriguing to consider how these factors combined in Jenny's case to provide the inspiration that created the concepts in *Factors of Soil Formation*. Hans Jenny arrived in the United States, in the fall of 1926, as a Rockefeller Fellow with an invitation to work in the laboratory of the future Nobel laureate Selman Waksman. While working in Waksman's lab, he participated in the group seminars. A recollection of one of these seminars (5) provides a glimpse that Jenny was already beginning to formulate a theoretical basis of soil formation. "... I gave a few interesting seminars. For example, in Zurich, we had a student who tried to correlate the European soil types with rainfall and evapotranspiration. Alfred Meyer was his name. He developed a sort of a scheme, a NS–quotient, that permitted correlating climatic moisture values with soils. So, while working under Waksman, I took that over and very boldly constructed a climatic moisture map of the United States saying, 'There is no law against doing it.' I predicted, for example, that [In] this place (e.g. Kansas) you must have a Chernozem. You must have a soil with a lime horizon.' Well, that impressed him. I didn't know if it was true. I said it must be so because I believed whole-heartedly in this climatic dependency, which was an oversimplification. But nevertheless, I brought these together and talked about humus, especially where much humus should be found according to climatic principles. Waksman was quite impressed with this. He always invited me home and took an interest in my private life, too."

Thus, with limited field experience, restricted entirely to the soils of Europe, Jenny began formulating a theory that allowed him to predict the distribution of soil properties based on external variables. The development of theory in the earth sciences, *a priori* to field observation, is probably not as rare as one might expect (7). What remained for Jenny was the opportunity to see the vastness of nature in person, and recognize the applicability of his evolving ideas.

This opportunity came soon. In the summer of 1927, Jenny was appointed as an official Swiss delegate to the First International Congress of Soil Science in Washington, D.C. The grand finale of this important event was a continental excursion, by train, to view the soils and landscapes of North America. For Jenny, and the effect it had on his ideas, this excursion must be compared in importance to Darwin's voyage on the *Beagle*. He later recounted:

"To me, the tour experience was a thrill and it opened a new world. I was impressed when we went from Washington south and saw the southern red soils and, a few weeks later, the black soils of Canada. I searched for a connection between the two. There must be a relationship, I thought, but in Europe the landscape is all cut up. The Alps wreck climatic gradients and smoothness of land, whereas in the wide plains of the United States there they were.

"The red soils of the South and the black soils of Canada were showcases of the climatic theory of soil formation. No doubt, our European textbooks by Ramann and Glinka were right in principle. I was intrigued by the 1,000 mile-long smooth transition of the Great Plains area, without interruption by massive mountains a la (the) Alps. The rolling plains, I fancied, must harbor the secret of mathematical soil functions. At times I could hardly sleep thinking about it."

The effect on his research was immediate. Following the excursion, Jenny was offered a chance to spend a year (which ultimately turned into a permanent position) at the University of Missouri at Columbia. Of this period, Jenny said, "Later, after the tour, I think that the lack of information on the connections (between soils across the continent) pushed me to try to find solutions, especially when I started teaching" (5). Beginning in 1928 (8), a series of research papers was published that searched for quantitative links between climate and various soil properties. However, the journey between these early papers and the detailed definitions and concepts presented in *Factors* was a long one. It included an exploration for other variables, years of detailed conceptual discussions with numbers of his colleagues, and a move to the University of California at Berkeley, where he realized his dream of filling the position first held by one of his intellectual inspirations, E. W. Hilgard.

The question might now be asked, "Why should one read a book that is now more than a half-century old?" In reply, I would respond with these arguments. First, *Factors of Soil Formation* is a theory of how soils and, although Jenny did not use the word, ecosystems form. Just because a theory is old, that does not mean it is out of date. While many advances have been made on *process* models of soil development, there have been no new factorial theories to replace that outlined in *Factors*. This should be of no great surprise, since the replacement of one scientific theory by another occurs so rarely, and has such a great significance, that it is said to constitute a scientific revolution (9). Secondly, for many of us, our introduction to Jenny's ideas has been presented to us through textbooks, many of which have oversimplified the vision contained in *Factors* or, in some cases, completely misrepresented it. Because the book has been out of print for much of the time since its first printing, the opportunity to read Jenny in its original form will prove to be an illuminating experience for many. Thirdly, the book is a pleasure to read. Its conciseness and clarity are matched by few books in the sciences. The numerous illustrations, which lend a pleasing visual quality to the book, were drawn primarily by Jenny himself. The book is packed with ideas that will stimulate any student of the earth today.

As for myself, I am greatly indebted to Dover Publications for this reprinting of *Factors of Soil Formation*. My copy, an original 1941 edition that was a gift from a retired colleague, was autographed by Jenny a few years before his death. During numerous rereadings of this book, I often wished to make notes, and highlight certain sections, but refrained because I did not want to mar what is, to me, a priceless work. However, with my new Dover edition, I promise, in future rereadings, to fill the margins with the ideas and thoughts that the text inspires. For I remain a student of Jenny.

RONALD AMUNDSON

BERKELEY, CALIFORNIA
December 1992

Literature Cited

1. DARWIN, C.: "The Origin of Species by Means of Natural Selection," John Murray, 1859.
2. JENNY, H.: "Factors of Soil Formation. A System of Quantitative Pedology," McGraw-Hill Book Company, Inc., New York, 1941.

3. AMUNDSON, R., and JENNY, H.: The place of humans in the state factor theory of ecosystems and their soils, *Soil Science*, 151: 99–109, 1991.
4. RETALLACK, G. J.: "Soils of the Past: An Introduction to Paleopedology," Unwin-Hyman, London, 1990.
5. JENNY, H.: "Hans Jenny. Soil Scientist, Teacher, and Scholar. Interviews by D. Maher and K. Stuart," *University History Series*, Regional Oral History Office, The Bancroft Library, University of California, Berkeley, 1989.
6. GOULD, S. J.: Darwin's Middle Road. In "The Panda's Thumb. More Reflections in Natural History," W. W. Norton and Company, New York, 1980, pp. 59–68.
7. GOULD, S. J.: "Time's Arrow, Time's Cycle. Myth and Metaphor in the Discovery of Geologic Time," Harvard University Press, Cambridge, Massachusetts, 1987.
8. JENNY, H.: Relation of climatic factors to the amount of nitrogen in soils. *J Am. Soc. Agron.*, 20: 900–912, 1928.
9. KUHN, T. S.: "The Structure of Scientific Revolutions. Second Edition," The University of Chicago Press, Chicago, 1970.

PREFACE

The College of Agriculture of the University of California offers a four-year curriculum in soil science. The first two years are devoted to the fundamental sciences, whereas the remaining period covers the field of soil science and related agricultural and scientific phases. Among the subjects prescribed, the four-unit course on "Development and Morphology of Soils" includes a study of soil-forming factors and processes of soil genesis. The present monograph is an extension of the first part of the course. The book must be classified as an advanced treatise on theoretical soil science.

Pedology is sometimes identified with the section of the domain of soil science that studies the soil body in its natural position. It is in this sense that the term is used throughout the book. As far as the author is aware the approach and presentation of the subject matter are entirely novel. They are the result of intensive research and a dozen years of teaching, beginning with an instructorship at the Federal Technical Institute in Zurich, Switzerland, followed by an association with the University of Missouri from 1927 to 1936.

It is impossible to acknowiedge adequately and specifically the assistance, criticisms, and encouragement rendered by scores of colleagues and students. To all of these the author tenders his sincere thanks. The author wishes to express his deep indebtedness to Dr. Roy Overstreet, who has given much time to long and profitable discussions. He has improved the manuscript logically and technically. In particular his contribution to the elucidation of the role of organisms in the scheme of soil formers will be appreciated by all who have been baffled by the complexity of the biotic factor. The author's profound thanks are also due to Dr. J. Kesseli of the Department of Geography, who read the manuscript and offered many helpful suggestions. The author extends his appreciation to Dr. R. H. Bray of the University of Illinois and to Dr. A. D. Ayers of the United States Salinity Laboratory for the use of unpublished data on loessial soils and on salinization. It is a pleasure to acknowledge the cooperation of members of the personnel of the Works Progress Administration Official Project No. 465-03-3-587-B-10, who assisted in the stenographic work and furnished translations from recent Russian literature.

The author wishes to add that the data selected from the literature are presented to illustrate pedological relationships. The selection does not reflect the author's opinion regarding the validity of these data nor does it indicate any discrimination against investigations that are not mentioned in the text.

HANS JENNY.

BERKELEY, CALIF.,
June, 1941.

INTRODUCTION

The vast importance of the soil in the development of various systems of agriculture and types of civilizations has long been recognized; but it is only within the last few decades that soils as such have been studied in a scientific manner. During thousands of years mankind has looked upon soils mainly from the utilitarian point of view. Today it is being realized more and more that the soil per se is worthy of scientific study, just as animals, plants, rocks, stars, etc., are subjects for theoretical research and thought. There is every reason to believe that any advance in the fundamental knowledge of soils will immediately fertilize and stimulate practical phases of soil investigations.

Since the beginning of the present century a great amount of work on soil identification and mapping has been carried out in all parts of the world. The detailed descriptions of the soil types investigated embrace hundreds of volumes, charts, and atlases. Attempts to coordinate the great mass of data frequently have been made, but almost exclusively along the lines of soil classification. The idea of classification has stood foremost in the minds of many great soil scientists of the past, and the present-day leaders in field soil studies continue in this same direction.

It should be remembered, however, that classification is not the only way to systematize facts. Data can also be organized by means of laws and theories. This method is characteristic of physics, chemistry, and certain branches of biology, the amazing achievements of which can be directly attributed to a great store of well-established numerical laws and quantitative theories. The present treatise on soils attempts to assemble soil data into a comprehensive scheme based on numerical relationships. Soil properties are correlated with independent variables commonly called "soil-forming factors." It is believed that such a mode of approach will assist in the understanding of soil differentiations and will help to explain the geographical distribution of soil types. The ultimate goal of functional analysis is the formulation of quantitative laws that permit mathematical treatment. As yet, no correlation between soil properties and conditioning factors has been found under field conditions which satisfies the requirements of generality and rigidity of natural laws. For this reason the less presumptuous name, "functional relationship," is chosen.

CONTENTS

FACTORS OF SOIL FORMATION

CHAPTER I

DEFINITIONS AND CONCEPTS

As a science grows, its underlying concepts change, although the words remain the same. The following sections will be devoted to an analysis of terms and concepts such as soil, environment, soil-forming factors, etc. The present method of treatment of soils is only one out of many, but it behooves a scientific system to be consistent in itself.

Preliminary Definitions of Soil.—In the layman's mind, the soil is a very concrete thing, namely, the "dirt" on the surface of the earth. To the soil scientist, or pedologist, the word "soil" conveys a somewhat different meaning, but no generally accepted definition exists.

Hilgard (4) defined soil as "the more or less loose and friable material in which, by means of their roots, plants may or do find a foothold and nourishment, as well as other conditions of growth." This is one of the many definitions that consider soil primarily as a means of plant production.

Ramann (7, 8) writes: "The soil is the upper weathering layer of the solid earth crust." This definition is scientific in the sense that no reference is made to crop production or to any other utilitarian motive.

Joffe (5), a representative of the Russian school of soil science, objects to Ramann's formulation on the grounds that it does not distinguish between soil and loose rock material. According to Joffe,

> The soil is a natural body, differentiated into horizons of mineral and organic constituents, usually unconsolidated, of variable depth, which differs from the parent material below in morphology, physical proper-

ties and constitution, chemical properties and composition, and biological characteristics.

It is problematic whether any definition of soil could be formulated to which everyone would agree. Fortunately there is no urgent need for universal agreement. For the purpose of presentation and discussion of the subject matter it is necessary only that the reader know what the author has in mind when he

FIG. 1.—Virgin prairie soil, Missouri. This soil profile shows a diffusion-like distribution of organic matter with depth. (*Courtesy of Soil Conservation Service.*)

uses the word "soil." This common ground will be prepared in the following sections.

The Soil Profile.—In order to gain a more concrete notion of the term "soil," the reader is directed to turn his attention to Figs. 1 and 2, which represent typical soils as found in the United States and other parts of the world. The pedologist's concept of soil is not that of a mere mass of inorganic and organic material; rather it takes cognizance of a certain element of organization that persistently presents itself in every soil. Although soils vary widely in their properties, they possess one common feature: they are anisotropic.

To a certain extent, many geological formations such as granite, loess, limestone, etc., are macroscopically *isotropic*, *i.e.*, the physical and chemical properties are independent of direction. If we draw a line through a huge block of granite, we find a certain sequence of quartz, feldspar, and mica, and of the elements

FIG. 2.—Podsol soil. This type of profile exhibits marked horizon differentiation. Organic matter is accumulated mainly on the surface. The white, bleached zone (A_2 horizon) is nearly free of humus. It overlies a dark brown layer of accumulations (B horizon) which contains moderate amounts of organic matter. (*From the late Prof. C. F. Shaw's collection of photographs.*)

silicon, aluminum, oxygen, etc. The same type of distribution pattern will be observed along any other line, selected in any direction (Fig. 3).

All soils are *anisotropic*. The spatial distribution of soil characteristics is not randomized but depends on direction. Along a line extending from the surface of the soil toward the center of the earth—arbitrarily denoted as Z-axis—the sequence of soil properties differs profoundly from that along lines parallel to the surface (Fig. 3). The soil has vectorial properties.

In the language of the pedologist, the anisotropism of soils is usually expressed with the words: "The soil has a profile." Its features are easily put into graphic form by choosing the vertical

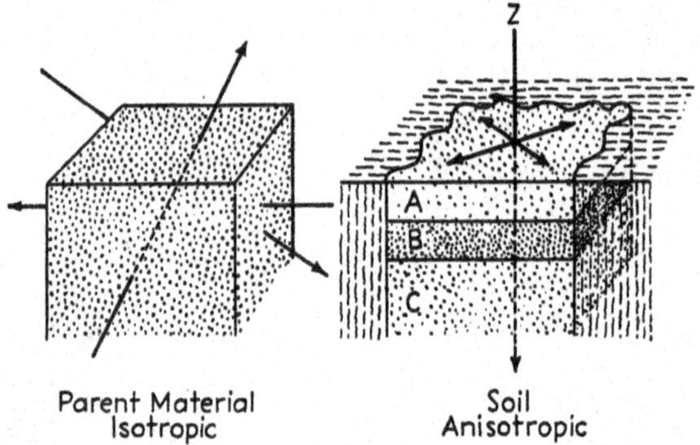

FIG. 3.—Illustrating an *isotropic* type of parent material and the *anisotropic* soil derived from it.

axis (Z-axis) as abscissa and plotting the soil properties on the ordinate, as shown in Fig. 4.

The curves in Fig. 4 exhibit well-defined hills or valleys, or relative maxima and minima. Pedologists call them "horizons."

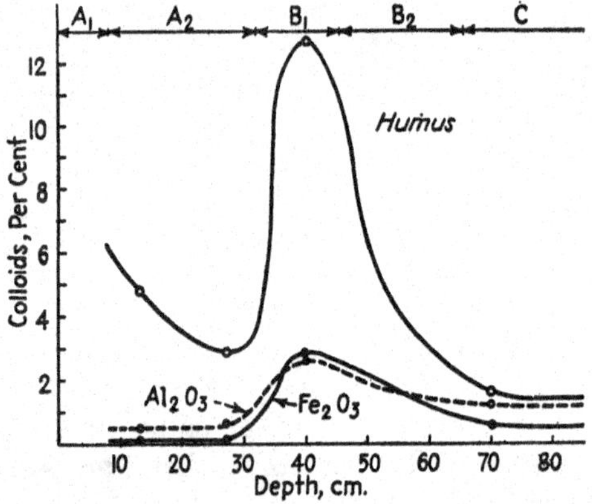

FIG. 4.—Three soil property-depth functions with maxima and minima (podsol profile). The ordinate indicates the amount of colloidal material in the various horizons.

In the field, those zones of abundances and deficiencies run approximately parallel to the surface of the land.

Naturally, every soil property has its own vertical distribution pattern or specific "depth function." In practice, special emphasis is placed on substances that migrate easily within the soil, such as soluble salts and colloidal particles. Their minima and maxima are named with capital letters *A*, referring to minima, *B*, referring to maxima, and *C*, which applies to the horizontal branch of the curve. The interpretation of the horizons is as follows:

Horizon *A*: *Eluvial horizon*, or leached horizon. Material has been removed from this zone.

Horizon *B*: *Illuvial horizon*, or accumulation horizon, in which substances, presumably from *A*, have been deposited.

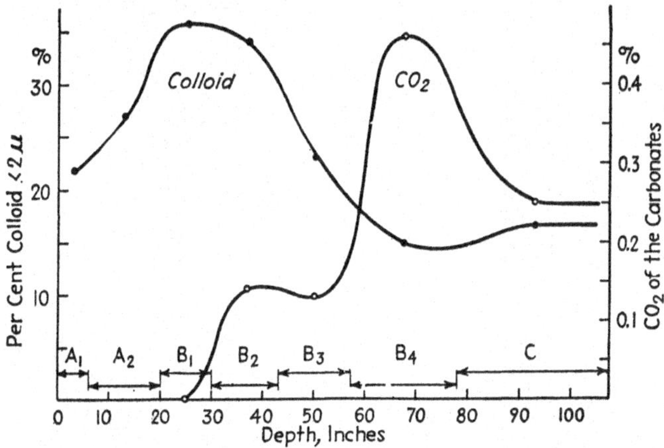

FIG. 5.—Colloidal clay and CO_2 of carbonates as a function of depth of a clay-pan soil. The maxima of these two soil properties do not occur at the same depth.

Horizon *C*: *Parent material*, from which the soil originated.

Frosterus (2) designates the zone of maxima and minima $(A + B)$ as *solum*. Variations within the horizons are indicated by subscripts, like A_1 and A_2, or, B_1, B_2, B_3, etc. Organic-matter deposits on top of the mineral soil are often labeled as A_o, *F*, *H*, etc.

If it so happens that several soil characteristics have maxima and minima that coincide spatially, as in Fig. 4 (podsol profile), the *ABC* terminology affords an easy means of describing and classifying soil profiles.

Not all soils possess such simple patterns. In the clay-pan soil shown in Fig. 5, the maximum for the carbon dioxide content (carbonates) occurs at greater depth than the peak of the colloid (clay particles). Assignment of the letters *A* and *B* is left to individual judgment. In the United States, the accumulation horizon frequently, but not exclusively, refers to the zone of enrichment in clay particles, and some uncertainty in horizon designations still exists.

The difficulty of horizon designation is accentuated in soils derived from anisotropic parent materials such as stratified sand and clay deposits. As a matter of fact, there exists great need for rigorous criteria of horizon identification, because all scientific

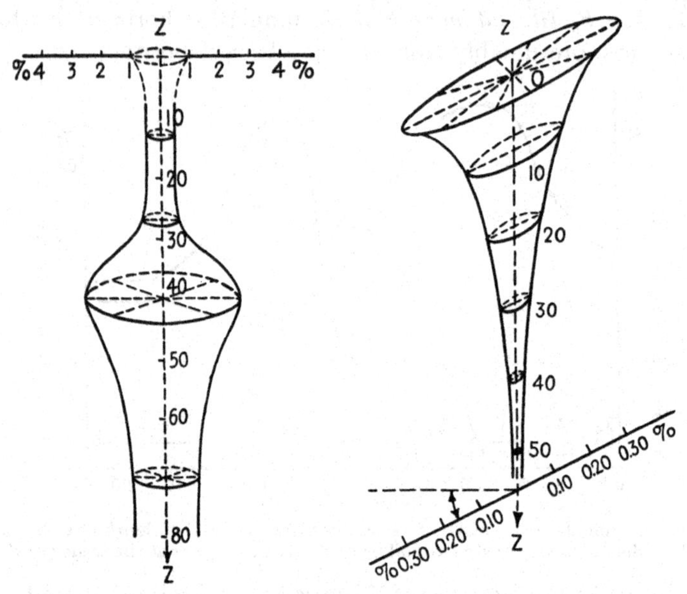

FIG. 6.—Illustration of a soil indicatrix. Colloidal Al_2O_3 of a podsol profile on level topography.

FIG. 7.—Illustration of a soil indicatrix. Total nitrogen of a prairie soil profile on a slope of 26°34′ or 50 per cent.

systems of soil classification as well as the theories regarding soil development rest on horizon interpretations.

The Soil Indicatrix.—On level land, soils derived from homogeneous, isotropic parent material are anisotropic only along the Z-axis. In the direction of right angles to it, the properties are isotropic. If the graph of Fig. 4 is rotated about the depth axis, a figure of revolution is obtained that indicates the spatial distribution of the properties of a soil. This figure might be designated as soil indicatrix. In the case of the distribution of colloidal alumina in Fig. 4, the indicatrix resembles a vase, as shown in

Fig. 6. If the surface of the land deviates from horizontal, the shape of the indicatrix becomes asymmetrical (Fig. 7) and is no longer a simple figure of revolution. The concept of the soil indicatrix offers a convenient tool for purposes of clarification and refinement of soil-profile descriptions.

Soil Defined as a System.—In this book, the soil is treated as a physical system. The word "physical" is inserted to distinguish it from purely logical systems, and the term corresponds with Joffe's statement that the soil is a natural body. The soil system is an open system; substances may be added to or removed from it.

Every system is characterized by properties that we may designate by symbols, such as s_1, s_2, s_3, s_4, s_5, etc. For example, s_1 may indicate nitrogen content, s_2 acidity, s_3 apparent density, s_4 amount of calcium, s_5 pressure of carbon dioxide, etc. Any system is defined when its properties are stated.

We further assume that the foregoing properties possess not only qualitative but also quantitative character, *i.e.*, we may express them with numerical figures. At present, not all soil properties have been sufficiently studied to permit quantitative expression, but there is reason to believe that the properties that can be given quantitative representation gradually will be increased in number as scientific research goes on.

We shall now make the additional obvious but important assumption that the properties of the soil system are functionally interrelated. That is to say, they stand in certain relationships to each other. If one property changes, many others also change. For instance, if water is added to a soil, not only is the magnitude of the moisture content increased but other properties like density, heat capacity, salt concentration, etc., are also altered. The simplest method of expressing the foregoing assumptions of interrelationships is in form of an equation, written as follows:

$$F(s_1, s_2, s_3, s_4, \cdots) = 0 \qquad (1)$$

It means that the soil is a system the properties of which are functionally related to each other.

Equation (1) is far too general to be of practical value to soil science. In fact, it is merely the definition of any natural body written in symbolic form. In order to distinguish soil from other natural systems, certain limits must be given to the properties

s_1, s_2, etc. The magnitudes of the properties must not exceed or fall below certain characteristic values. To quote an example, the water content of a soil must be below, say, 95 per cent on a moist basis, or else the system would not be called "soil" but "swamp," or "lake," or "river."

No general attempt as yet has been made to assign quantitative limits to soil properties, and, therefore, it is not possible at the present time to contrast soils sharply with other natural bodies on the basis of Eq. (1). However, if we specify for the moment that soils are portions of the upper weathering layer of the solid crust of the earth, the limits of some of the properties of Eq. (1) are roughly set.

Soil and Environment.—The latter contention requires a closer examination. Just where on the surface of the earth are the soils proper, and what constitutes the boundaries between soils and other natural bodies that also are part of the upper portion of the earth?

It is generally realized that the soil system is only a part of a much larger system that is composed of the upper part of the lithosphere, the lower part of the atmosphere, and a considerable part of the biosphere. This larger system is illustrated in Fig. 8.

A number of Russian soil scientists designate this larger system as soil type (climatic soil type). Glinka (3) writes:

> Contrary to the majority of foreign soil scientists, Russian pedologists choose the soil type as a unit of classification instead of soil masses, regarding the soil type as a summary of the external and internal properties of a soil. Thus, speaking of chernozem, for instance, the Russian pedologist saw not only a natural body with definite properties, but also its geographical position and surroundings, *i.e.*, climate, vegetation, and animal life.

This Russian idea of soil type is nearly identical with the soil geographer's concept of natural landscape.

Most soil scientists deal only with part of this wider system, namely, the soil per se. The remaining part is called "environment." Often it is not sufficiently realized that the boundary between soil and environment is artificial and that no two soil scientists have exactly the same enclosure of the soil system in mind. A glance at Fig. 8 provides a felicitous illustration. Suppose we approach the surface from the atmosphere, just where does the soil begin? Is the forest litter a part of the soil or of the surroundings? If one is inclined to discard the freshly fallen leaves or needles and include only the decomposed organic material in the soil system, what degree of decomposition is necessary? Similar difficulties are encountered in the sampling of virgin prairie soils. Living grasses are not soil, but dead parts gradually become incorporated into the soil matrix. In practice,

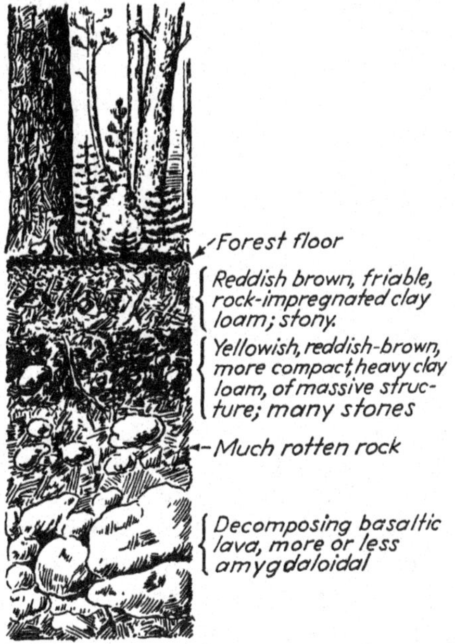

FIG. 8.—The larger system. (*After a drawing by G. B. Bodman.*)

a rather arbitrary separation between vegetative cover and soil becomes necessary.

If we approach soils from below, similar problems are encountered. There is no sharp boundary between undecomposed rock, weathered rock, soil material, and soil. Although criteria for distinguishing soils from unconsolidated geological deposits may be formulated (see page 17), they are of necessity artificial. In the opinion of the author, the distinction between soil and environment is arbitrary; it exists only in our minds, not in nature. The often quoted axiom that soils are "independent natural bodies" is misleading, and little is gained by trying to establish tight compartments between pedology and related sciences.

States of the Soil System.—A system is said to assume different states when one or more of its properties undergo a change. Returning to Eq. (1)

$$F(s_1, s_2, s_3, s_4, s_5, \cdots) = 0 \tag{1}$$

an increase in, say, the nitrogen content (s_1) produces a different state, or a different soil. Theoretically speaking, the smallest change in any one of the properties, denoted by the differential ds, gives rise to a new soil. For practical purposes, different states, *i.e.*, different soils, are only recognized when the properties are changed to such an extent that the differences may be easily ascertained by mere field inspection. In the American Soil Survey, the various states of the soil system are known as "soil types."* The number of recognized soil types in the United States amounts to several thousands and is increasing steadily. As new areas are surveyed, new soil types are discovered. Remapping of previously surveyed regions often results in refinements with a corresponding splitting of existing soil types into two or more new ones. On theoretical grounds, it follows from Eq. (1) that the number of soil states and, consequently soil types is infinite.

Some Especially Important Soil Properties.—We have previously stated that s_1, s_2, s_3, etc., represent soil properties, and there is general agreement that nitrogen, acidity, color, etc., are typical soil characteristics. There are, however, a number of properties of the soil system that are not universally recognized as soil properties. They are the following: soil climate (soil moisture, soil temperature, etc.), kind and number of soil organisms, and topography, or the shape of the surface of the soil system. These properties will be denoted by special symbols (cl' = climate, o' = organisms, r' = topography or relief), which are included in Eq. (1)

$$F(cl', o', r', s_1, s_2, s_3, \cdots) = 0 \tag{2}$$

* According to Marbut (6) a soil type, as the term is used in the United States, is a soil unit based on consideration of all soil characteristics and is designated by the series name and texture description, *e.g.*, Norfolk sandy loam.

There is no essential difference between Eqs. (1) and (2) except that some of the soil properties have been grouped together and given special symbols. The reason for doing so will become obvious at a later stage of our discussion. Emphasis should be placed on the fact that in Eq. (2) the soil system is defined or described by its own properties and nothing else. Moreover, soil is treated as a static system. No reference is made that the properties may change with time.

Soil Formation.—The transformation of rock into soil is designated as soil formation. The rock may be gneiss, limestone, shale, sand, loess, peat, etc. To avoid too liberal an interpretation of the term "rock," soil scientists prefer to use the expression "parent" material or "soil" material. The relationship between parent material, soil formation, and soil may be conveniently expressed as follows:

$$\text{Parent material} \xrightarrow[\text{Soil formation}]{} \text{Soil}$$

The foregoing formulation introduces a new factor or variable into our discussion, namely, *time*. The states of the soil system vary with time, *i.e.*, they are not stable. Suppose we consider a piece of granite that is brought to the surface of the earth. In the interior of the earth, the granite may have been in equilibrium with its immediate surroundings; but now, on the surface of the earth, it is in an entirely new environment, and the rock system is highly unstable. It is continuously changing its properties in a definite direction, namely, toward a new equilibrium state. When the final equilibrium state has been reached, the process of transformation, of soil formation, has been completed, and the rock has become a *mature soil.* It is customary to designate the intermediate, unstable states as *immature soils.* We may define the phases of soil formation as follows:

$$\underset{\substack{\text{Initial state}\\ \text{of system}}}{\text{Parent material}} \xrightarrow[\text{Intermediate states}]{} \underset{\substack{\text{Final state}\\ \text{of system}}}{\text{Soil (mature)}}$$

In this formulation, soil is treated as a dynamic system. Emphasis is placed on the changes of the properties of the soil as a function of time.

It may be well to point out that the foregoing concept of soil formation is broader than that of a certain group of soil scientists who sharply distinguish between weathering and soil formation. The former process is said to be geologic and destructive, whereas the latter is pedologic and creative. In the present treatise, we adopt a more conservative viewpoint and consider weathering as one of the many processes of soil formation.

Soil-forming Factors.—Agriculturists have long realized that many important properties of soils are inherited from the underlying rocks. Technical expressions like limestone soils or granitic soils are encountered in the oldest textbooks on agricultural subjects. They clearly convey the importance of parent material in soil formation.

However, it remained for Hilgard in America and, independently, for Dokuchaiev in Russia to enunciate the important discovery that a given parent material may form different soils depending on environmental conditions, particularly climate and vegetation.

Parent material, climate, and organisms are commonly designated as soil formers or soil-forming factors. Since soils change with time and undergo a process of evolution, the factor time also is frequently given the status of a soil-forming factor. Topography, which modifies the water relationships in soils and to a considerable extent influences soil erosion, also is usually treated as a soil former.

In view of our discussion on soil as a system and its relation to environment, the question immediately arises as to what the precise nature of soil-forming factors really is. Are they soil properties or environmental factors or something entirely different? Air climate undoubtedly is a property of the environment. As regards organisms, some belong to the environment (*e.g.*, trees); others are wholly within the soil (*e.g.*, protozoa). Topography belongs to both soil and environment, time to neither of them. What, then, may we ask, is the fundamental feature common to these factors that has induced soil scientists to assemble them into a distinguished class, the "soil formers"?

Joffe identifies two kinds of soil formers. passive and active. He defines:

The *passive* soil formers are represented by the constituents that serve as the source of the mass only and by the conditions that affect the mass. They comprise the parent material, the topography, and the age of the land. The *active* soil formers are the agents that supply the energy that acts upon the mass furnishing reagents for the process of soil formation. The elements of the biosphere, the atmosphere, and partly the hydrosphere are representative of this class of soil formers.

Joffe's contrasting of mass and energy is appealing as regards parent material and climate, but the role attributed to the factors time and topography leads to confusion.

Vilensky and the Russian scholars in general identify soil-forming factors with outward factors. Marbut also speaks of environmental factors, and it seems that he uses soil-forming factors and environment synonymously. However, the word "environment" would be stretched beyond its common meaning if one were to consider the microorganisms living within the soil as environment.

Others, like Glinka, apply the word "forces," but in a mystic rather than a physical sense. These forces are not amenable to quantitative elucidation.

Another group of pedologists calls soil-forming factors the causes and soil properties their effects. These scientists operate with the causality principle of the nineteenth century philosophers (*e.g.*, Mill). The introduction of causality aspects to soil formation is not fruitful. It unnecessarily complicates matters, because every soil property may be considered a cause as well as an effect. For example, soil acidity influences bacteria and thus acts as a cause. On the other hand, bacteria may change the acidity of a soil, which then assumes the status of an effect. Again, the factor time does not fit into the causality scheme, since time itself can be neither cause nor effect.

To summarize, we come to the conclusion that no satisfactory and consistent definition of soil formers exists.

A New Concept of Soil-forming Factors.—Soil is an exceedingly complex system possessing of a great number of properties. One might contend that a soil is defined only if all its properties are explicitly stated. Fortunately, there are reasons to believe that such a Herculean task is not required. According to Eq. (2) the properties of a soil are functionally interrelated; therefore, if a sufficient number of them is fixed, all others are fixed. From investigations of systems simpler than soils, we know that a limited number of properties will suffice to determine the state of a system. If we have, for example, 1 mol of oxygen gas and know its temperature and pressure, then a great number of other properties of the gas, like density, average velocity of the molecules, heat capacity, etc., are invariably fixed. The properties capable of determining a system are known as "conditioning" factors. Their nature is such that they can be made to vary independently of each other. They are independent variables.

In reference to soils, two questions present themselves: what are the conditioning factors, and what is the minimum number necessary to define completely the soil system? A priori we do not know. Experience has shown, however, that some soil properties satisfy the requirements of an independent variable, whereas others do not. With reference to the latter, it is evident that the hydrogen ion concentration (acidity) and the hydroxyl ion concentration (basicity) cannot be selected as a pair of independent variables, because a change in one necessitates a change in the other. They cannot be made to vary independently of each other. Similarly, soil structure and organic matter or soil color and ferric oxide content are properties that often change simultaneously. A different situation exists in the case of soil temperature and soil moisture. Soils may possess high temperatures and at the same time low moistures, and vice versa. One may change without altering the other. These two soil properties are independent variables. Likewise, the shape of the surface of the soil, *i.e.*, the topography, belongs in this class, as do certain aspects of the organisms.* These soil properties, or groups of properties, soil climate, organisms, and topography, are listed in Eq. (2) as cl', o', and r'. There remain in Eq. (2) a great number of s values, s_1, s_2, s_3, etc. In view of their great number and variety, it is not to be expected that they are conditioned solely by the three variables cl', o', and r'. Unfortunately, we do not know at the present stage of soil science what group of s values can be treated as independent variables. An additional approach must be sought. It is found by considering soil as a dynamic system.

If we admit that soil formation consists of a series of chemical and biochemical processes, we may again resort to analogies with simpler systems. Dynamics of chemical reactions may be described accurately by indicating the initial state of the system, the reaction time, and conditioning variables. These considerations may be directly applied to the soil. The initial state of the

* For details compare with p. 199.

soil system has been designated on page 11 as parent material. Since reaction time and time of soil formation are analogous, we may include the two independent variables parent material and time with the conditioning parameters and postulate that the following factors completely describe the soil system:

Independent variables or soil-forming factors		
	Climate	(cl')
	Organisms	(o')
	Topography	(r')
	Parent material	(p)
	Time	(t)

These terms are identical with the soil formers previously mentioned, but their meaning is different. They are not forces, causes, or energies, nor are they necessarily environment. They have but one feature in common. They are the *independent variables that define the soil system.* That is to say, for a given combination of cl', o', r', p, and t, the state of the soil system is fixed; only one type of soil exists under these conditions.

In this new interpretation of soil-forming factors, the notions of "forming" or "acting" that connote causal relationships have been replaced by the less ambiguous conceptions of "defining" or "describing." We realize, of course, that for everyday usage it may be convenient to think of some of the soil-forming factors as "creators." However, as soon as one undertakes to examine all factors more critically, the causality viewpoint leads into so many logical entanglements that it appears preferable to drop it altogether and adopt the descriptive attitude outlined above.

Relationship between Soil Properties and Soil-forming Factors.—Since the soil-forming factors completely define the soil system, all s values must depend on cl', o', r', p, and t, a dependency that may be expressed as

$$s = f'(cl', o', r', p, t) \tag{3}$$

This equation states that the magnitude of any one of the properties of the s type such as pH, clay content, porosity, density, carbonates, etc., is determined by the soil-forming factors listed within parentheses. The letter f' stands for "function of," or "dependent on."

The Fundamental Equation of Soil-forming Factors.—All preceding discussions were restricted to the characteristics of the soil per se. Particularly the properties cl', o', and r', referred to those of the soil. In view of our discussion on the relationship between soil and environment (page 8), an additional formulation of the soil-forming-factor equation suggests itself.

Soil and environment form coupled systems. That is to say, many corresponding properties of the two systems pass continuously from the one to the other. They step across the boundaries. Temperature, for example, does not change abruptly as one passes from the soil to the environment. Neither do nitrogen, oxygen, and carbon dioxide content. Many organisms, especially vegetation, likewise are common to both soil and environment. It is well known that the root hairs of a tree are in intimate contact with the mineral particles and, in practice, are treated as soil properties. Similar treatment is accorded to the fine rootlets. At some point, the root system of the tree emerges into the trunk, and the latter is usually considered a part of the environment. Topography, *i.e.*, the shape of the upper boundary of the soil system, naturally is a property of both soil and environment.

We note, therefore, that the soil properties cl', o', and r' cross the boundaries of the soil system and extend into the environment. The concept of coupled systems suggests that we may replace the soil properties cl', o', and r' of Eq. (3) by their counterparts in the environment and thus obtain an *environmental* formula of soil-forming factors

$$s = f(cl, o, r, p, t, \cdots) \tag{4}$$

which we shall designate as the *fundamental* equation of soil-forming factors. It is identical with Eq. (3) except that the symbols cl, o, and r refer now to environment. The corresponding factors cl' and cl, o' and o, r' and r are assumed to be functionally interrelated. The dots indicate that, besides the variables listed, additional soil formers may have to be included in Eq. (4).

In selecting cl, o, r, p, and t as the independent variables of the soil system, we do not assert that these factors never enter functional relationships among themselves. We place emphasis on the fact that the soil formers *may* vary independently and *may* be obtained in a great variety of constellations, either in nature or under experimental conditions. It is well known that various kinds of parent materials and topographies do occur in various kinds of climates and that given amounts of annual precipitations are found in association with either low or high annual temperatures, and vice versa.

Soil Defined in Terms of the Fundamental Equation.—We are now in a position to establish a differentiation—arbitrary, to be sure—between soil and geological material. Soils are those portions of the solid crust of the earth *the properties of which vary with soil-forming factors,* as formulated by Eq. (4). As pedologists, we are interested only in those strata on the solid surface of the earth the properties of which are influenced by climate, organisms, etc. From this definition, it follows at once that the *depth of soils* is a function of soil-forming factors; in particular, it varies with humidity and temperature.

The Solution of the Fundamental Equation of Soil-forming Factors.—Half a century ago, Hilgard had recognized the existence of the soil-forming factors and discussed them at length in his classic book on soils. Dokuchaiev (1) likewise realized the existence of soil formers. He went a step further than Hilgard and formulated an expression somewhat similar to Eq. (4). However, he did not solve it. He wrote:

> In the first place we have to deal here with a great complexity of conditions affecting soil; secondly, these conditions have no absolute value, and, therefore, it is very difficult to express them by means of figures; finally, we possess very few data with regard to some factors, and none whatever with regard to others. *Nevertheless, we may hope that all these difficulties will be overcome with time, and then soil science will truly become a pure science.*

In these phrases Dokuchaiev expressed prophetic insight into one of the most fundamental problems of theoretical soil science, namely, the quantitative solution of the soil-forming-factor equation. Curiously enough, Dokuchaiev's students have paid little attention to the plea of their master for solving the function of soil-forming factors. Russian pedology and international soil science have developed in an entirely different direction and are stressing the subject of classification of soils.

The fundamental equation of soil formation (4) is of little value unless it is solved. The indeterminate function f must be replaced by some specific quantitative relationship. It is the purpose of the present book to assemble known correlations between soil properties and soil-forming factors and, as far as is possible, to express them as quantitative relationships or functions.

Generally speaking, there are two principal methods by which a solution of Eq. (4) may be accomplished: first, in a theoretical manner, by logical deductions from certain premises, and, second, empirically by either experimentation or field observation. At the present youthful stage of soil science, only the observational method—fortified by laboratory analyses of soil samples—can be trusted, and it must be given preference over the theoretical alternative.

The solutions given in this study are formalistic, as contrasted with mechanistic. In simple words, we endeavor to determine *how* soil properties vary with soil-forming factors. We shall exhibit but little curiosity regarding the molecular mechanism of soil formation and thus avoid lengthy excursions into colloid chemistry, microbiology, etc. Such treatment will be reserved for a later occasion.

Isolating the Variables.—Considerable controversy exists among soil scientists as to the relative importance of the various soil-forming factors. Climate is usually considered the dominant factor, but parent material still claims an impressive number of adherents. In recent years, the role of vegetation has come into the limelight.

In the main, these various claims are speculations rather than scientific facts, since no systematic quantitative study of the relationships between soil properties and all soil-forming factors has ever been made. To ascertain the role played by each soil-forming factor, it is necessary that all the remaining factors be kept constant. On the basis of the fundamental Eq. (4), we obtain the following set of individual equations of soil-forming factors:

$$s = f_{cl}\ (\text{climate})_{o,r,p,t\ldots}$$
$$s = f_{o}\ (\text{organisms})_{cl,r,p,t\ldots}$$
$$s = f_{r}\ (\text{topography})_{cl,o,p,t\ldots}$$
$$s = f_{p}\ (\text{parent material})_{cl,o,r,t\ldots}$$
$$s = f_{t}\ (\text{time})_{cl,o,r,p\ldots}$$

The subscripts indicate that the remaining soil-forming factors do not vary. For example, in order to study accurately the soil-climate relationships, it is necessary that comparisons be restricted to soils of identical origin and time of soil formation, etc.

From Eq. (4), it follows that the total change of any soil property depends on all the changes of the soil-forming factors, or, written in mathematical language

$$ds = \left(\frac{\partial s}{\partial cl}\right)_{o,r,p,t} dcl + \left(\frac{\partial s}{\partial o}\right)_{cl,r,p,t} do + \left(\frac{\partial s}{\partial r}\right)_{cl,o,p,t} dr + \left(\frac{\partial s}{\partial p}\right)_{cl,o,r,t} dp + \left(\frac{\partial s}{\partial t}\right)_{cl,o,r,p} dt \quad (5)$$

The quotients in parentheses are the partial derivatives of any soil property (s_1, s_2, etc.) with respect to the soil formers. Their numerical magnitudes are true indexes of the relative importance of the various soil-forming factors.

Some Inherent Difficulties.—Each soil-forming factor has been treated as a variable, which, broadly speaking, denotes anything that varies. In formulating Eq. (2) and especially Eq. (4), the concept of a variable has been considerably refined; in particular, it is now taken for granted that a variable may be expressed quantitatively, *i.e.*, by numbers. Little difficulty is encountered in assigning numbers to the variables time and topography. Climate cannot be described by a single index, but it may be split into separate factors each of which permits quantitative characterization. Parent material and organisms offer greater obstacles. At present, we cannot establish functional relationships between a soil property and various types of rocks or vegetational complexes, but we may at least compare soil formation on various substrata and under the influence of various types of vegetation and thus arrive at a quantitative grouping of these variables.

A serious practical difficulty in solving Eq. (4) in the field arises from the requirement of keeping the soil formers constant. In laboratory experiments on soil formation, we can exercise rigid control of the conditioning variables (*e.g.*, temperature, moisture, etc.) and thus obtain sets of data that leave no doubt as to the functional relationship between them. Under field conditions, considerable variation in the magnitude of the variables cannot be avoided, in consequence of which we arrive at scatter diagrams rather than perfect functions. Statistical considerations must be introduced, and the resulting equations possess the character of general trends only. Even so, the gain in scientific knowledge fully justifies the mode of approach.

Literature Cited

1. AFANASIEV, J. N.: The classification problem in Russian soil science, *U.S.S.R. Acad. Sci. Russian Pedological Investigations*, V, 1927.
2. FROSTERUS, B.: Die Klassifikation der Böden und Bodenarten Finnlands, *Mémoires sur la classification et la nomenclature des sols*, 141–176, Helsinki, 1924.
3. GLINKA, K. D.: Dokuchaiev's ideas in the development of pedology and cognate sciences, *U.S.S.R. Acad. Sci. Russian Pedological Investigations*, I, 1927.
4. HILGARD, E. W.: "Soils," The Macmillan Company, New York, 1914.
5. JOFFE, J. S.: "Pedology," Rutgers University Press, New Brunswick, N. J., 1936.
6. MARBUT, C. F.: Soils of the United States, *Atlas of American Agriculture*, Part III, U. S. Government Printing Office, Washington, D. C., 1935.
7. RAMANN, E.: "Bodenkunde," Verlag Julius Springer, Berlin, 1911.
8. RAMANN, E.: "The Evolution and Classification of Soils," W. Heffer & Sons, Ltd., London, 1928.

CHAPTER II

METHODS OF PRESENTATION OF SOIL DATA

Descriptions of soils by observers in the field are primarily of a qualitative character. Functional analysis of soils requires quantitative data. At present, these are mainly available in the domain of soil physics and soil chemistry. It is hoped that, in the future, soil surveyors also will stress the accumulation of quantitative data such as measurements of variability of horizons, topography, etc. It is desirable that the information be arranged in a manner suitable for presentation, if possible, in graphic form.

Presentation of Physical Analyses.—The list of measurable physical soil properties includes the true and apparent densities, heat capacity and conductivity, plasticity, soil-structure criteria, and a host of properties pertaining to water relationships, such as moisture equivalent, wilting percentage, water-holding capacity, vapor-pressure curves, etc. The variations of these properties with depth are conveniently presented by means of Cartesian coordinates and the soil-property indicatrix discussed on page 6.

A special problem is presented by the interpretation of mechanical analyses of soils, *i.e.*, the separations of soil particles—either ultimate or aggregates—into various size groups. Table 1 shows the designations and size limits of the American and the international systems of particle-size classification. Bradfield (9) and his pupils have developed methods for subdividing the clay fraction. Marshall (6) has been able to separate colloidal clay particles as small as 10 millimicrons (10mμ).

The relationship between particle size (upper limit) and settling velocity in water at 18°C. is calculated with the aid of Stokes' law (1, 8):

$$v = 34{,}760r^2$$

v denotes the settling velocity expressed in centimeters per second, and r the radius of the particle in centimeters. Many soil scientists express mechanical analyses exclusively in terms of settling velocities or their logarithms; others prefer to deal with

TABLE 1.—CLASSIFICATION OF PARTICLE SIZES
(U. S. systems and international system)

Designation	Diameter, millimeters	Logarithm of diameter, upper limit	Logarithm of settling velocity (v)
Fine gravel	2–1	0.301	2.541
Coarse sand	1–0.5	0.000	1.939
International coarse sand	*2–0.2*	*0.301*	*2.541*
Medium sand	0.5–0.25	−0.301	1.337
Fine sand	0.25–0.1	−0.602	0.735
International fine sand	*0.2–0.02*	*−0.699*	*0.541*
Very fine sand	0.1–0.05	−1.000	−0.061
Silt, prior to 1938	0.05–0.005	−1.301	−0.663
Silt	0.05–0.002	−1.301	−0.663
International silt	*0.02–0.002*	*−1.699*	*−1.459*
Clay, prior to 1938	<0.005	−2.301	−2.663
Int. clay and U. S. clay	*<0.002*	*−2.699*	*−3.459*

diameters of hypothetical or "effective" spherical particles computed from Stokes' law (Table 1).

Table 2 contains information with reference to the mechanical composition of a Cecil fine sandy loam from Rutherfordton, N. C.

The column labeled Summation percentage indicates the accumulated percentage of the fractions, starting with the

TABLE 2.—MECHANICAL ANALYSES OF CECIL FINE SANDY LOAM (*Atlas of American Agriculture, Part* III, *Soils of the United States*, p. 54)

Fractions	Upper limit, millimeters	*A* horizon 0–5 in.		*B* horizon 5–36 in.		*C* horizon 72–96 in.	
		Per cent	Summation percentage	Per cent	Summation percentage	Per cent	Summation percentage
Fine gravel (2–1 mm.)	2	4.8	100.5	2.8	100.0	1.8	100.3
Coarse sand (1–0.5 mm.)	1	12.4	95.7	4.6	97.2	12.7	98.5
Medium sand (0.5–0.25 mm.)	0.5	7.8	83.3	2.8	92.6	11.7	85.8
Fine sand (0.25–0.1 mm.)	0.25	27.2	75.5	11.7	89.8	37.7	74.1
Very fine sand (0.1–0.05 mm.)	0.1	9.8	48.3	5.5	78.1	8.0	36.4
Silt (0.05–0.005 mm.)	0.05	22.9	38.5	21.4	72.6	14.2	28.4
Clay (<0.005 mm.)	0.005	15.6	15.6	51.2	51.2	14.2	14.2

smallest fraction. For example, in the *A* horizon there are 48.3 per cent particles smaller than very fine sand. Figure 9 shows the summation percentages plotted against the particle

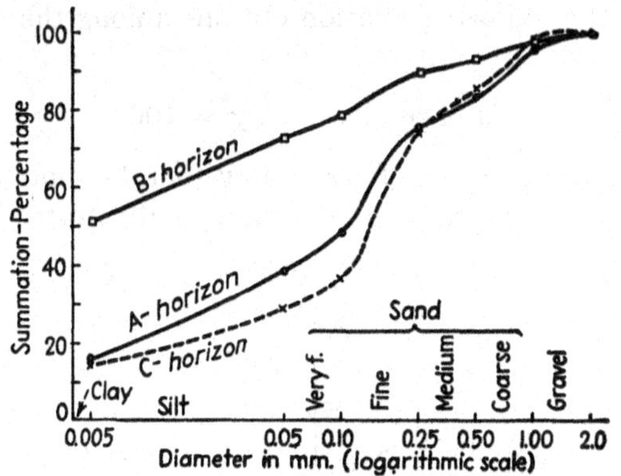

FIG. 9.—Graphic representation of the physical composition of Cecil fine sandy loam (Table 2).

size. Owing to the wide ranges of diameters, the fractions are indicated on a logarithmic scale (Table 1). If the summation percentages are plotted in relation to the logarithms of the settling velocities, a similar family of curves is obtained. A different method of presentation is illustrated in Fig. 10. Here the summation percentages are shown as a function of depth, a relationship that places emphasis on the profile distribution of the mechanical composition of the soil.

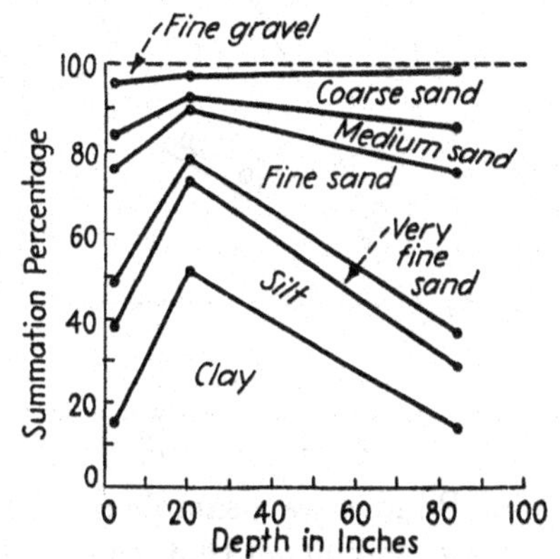

FIG. 10.—Graphic representation of the texture of Cecil fine sandy loam (Table 2). Emphasis is placed on the distribution of the fractions in relation to depth.

Colloid chemists favor size-distribution curves. In principle, they are obtained from the first derivatives of the summation-percentage curves. The magnitudes of the fractions appear as areas. In spite of certain advantages, this type of presentation has not been popular among soil scientists, partly because of the necessity of assuming a lower limit for the clay fraction.

Triangular coordinates provide a convenient method for condensing bulky tables on texture relationships. All coarse fractions are combined as "sand" and are contrasted with silt and clay. The following relation obtains among the percentage values:

$$\text{Sand} + \text{silt} + \text{clay} = 100$$

Each mechanical analysis of a soil may then be represented as a single point in a concentration diagram, as illustrated in Fig. 11.

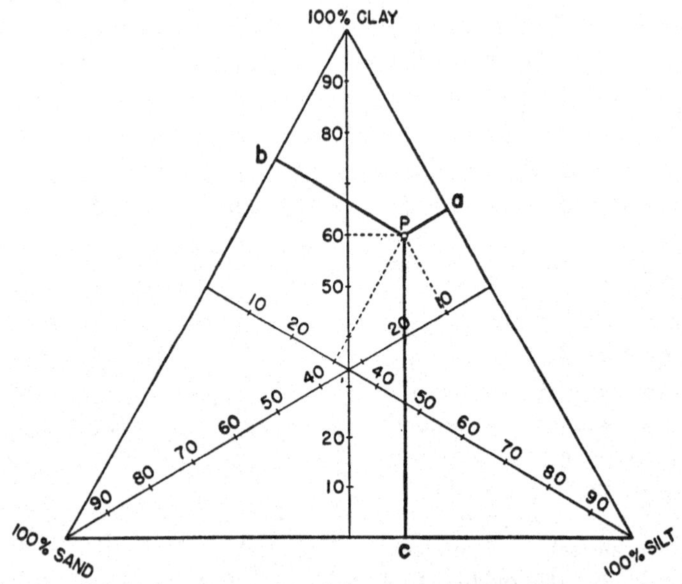

FIG. 11.—Triangular presentation of the physical composition of a soil. The point P corresponds to a soil which contains 10 per cent sand, 30 per cent silt, and 60 per cent clay. Note: $P \to a$ = sand, $P \to b$ = silt, and $P \to c$ = clay.

Point P indicates a sample consisting of 10 per cent sand, 30 per cent silt, and 60 per cent clay. The data for the Cecil fine sandy loam of Table 2 are depicted in Fig. 12. Instead of obtaining the percentage values from the projections upon the altitude of the triangle (Fig. 11), the data may be read off on its sides.

Interpretation of Chemical Analyses.—A detailed fusion analysis of a soil comprises the determination of over 20 individual constituents. For many practical purposes, this number is reduced to about a dozen substances: SiO_2, Al_2O_3, Fe_2O_3, CaO, MgO, K_2O, Na_2O, SO_4, P_2O_5, H_2O, organic matter, etc. If four horizons of a soil profile are analyzed, we are confronted with some 50 figures, and the handling and interpretation of such an array of data are as difficult a task as the analytical procedure itself.

Frequently fusion data are supplemented by analyses of HCl and H_2SO_4 extracts. In recent years, colloid chemical investigations have come into prominence (4). These include estimates of the colloid content of the soil, determination of adsorption

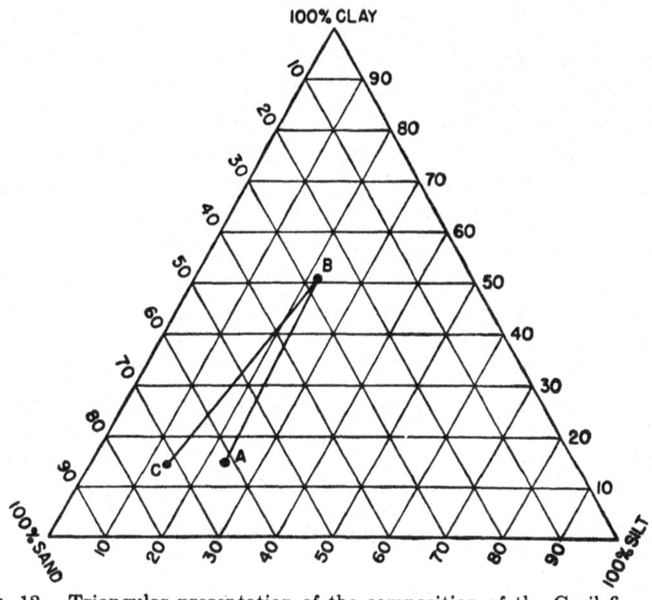

Fig. 12.—Triangular presentation of the composition of the Cecil fine sandy loam profile given in Table 2. Note the accumulation of clay in the *B* horizon.

capacities (base exchange capacities), exchangeable ions, electric potentials, etc.

For the purpose of interpretation of chemical data, it is advisable to construct a table of *molecular values* by dividing the customary percentage data by the molecular weights. If a soil contains 55.90 per cent SiO_2, the molecular value is $55.90/60 = 0.932$. Chemists are determining atomic weights more and more accurately, and this necessitates endless recalculations of molecular values. Since these changes are beyond the accuracy of soil and rock analyses, it appears expedient to use the rounded figures of H. S. Washington that Niggli (7) advocates for international use.

SiO_2 = 60	TiO_2 = 80	S = 32
Al_2O_3 = 102	P_2O_5 = 142	Cr_2O_3 = 152
Fe_2O_3 = 160	MnO = 71	NiO = 75
FeO = 72	ZrO_2 = 123	CoO = 75
MgO = 40	CO_2 = 44	BaO = 153.5
CaO = 56	SO_3 = 80	SrO = 103.5
Na_2O = 62	Cl_2 = 71	Li_2O = 30
K_2O = 94	F_2 = 38	

Molecular values offer many advantages. In the first place, we are not so much interested in the weight changes of the soil constituents as in changes in their atomic and molecular proportions. Stoichiometric relationships are more clearly brought out by molecular data than by weight figures. In the second place, chemistry has shown that chemical laws assume the simplest form when expressed in molecular relationships.

In order to reduce the number of items in a table of analyses, two or more values may be combined into ratios. The following quotients and symbols are often encountered in soil literature:

$$\frac{SiO_2}{Al_2O_3} = ki \text{ value [Harrassowitz (2)]} = sa \text{ value [Marbut (5)]}$$

$$\frac{SiO_2}{Fe_2O_3} = sf \text{ value (Marbut)}$$

$$\frac{SiO_2}{Al_2O_3 + Fe_2O_3} = \text{silica-sesquioxide ratio}$$

$$\frac{K_2O + Na_2O + CaO}{Al_2O_3} = ba \text{ value (Harrassowitz)}$$

$$\frac{K_2O + Na_2O}{Al_2O_3} = ba_1 \text{ value;} \qquad \frac{CaO + MgO}{Al_2O_3} = ba_2 \text{ value}$$

$$\frac{Al_2O_3}{Fe_2O_3}, \frac{K_2O}{Na_2O}, \frac{CaO}{MgO}$$

These ratios afford a means of detecting relative translocations of elements. If one assumes, for example, that Al_2O_3 is the most stable of the aforementioned compounds—because of its insolubility at neutral reaction—the ratios involving aluminum represent numerical indexes of relative accumulations and depletions in various horizons. The ratios may also serve as checks against faulty conclusions regarding actual losses of substances. It is frequently found that surface soils contain less silica (on a percentage basis) than the parent material, and one might conclude that silica has been leached out. This loss may be only fictitious, caused by addition of organic matter to the soil, which automatically lowers the percentage composition of SiO_2. An inspection of the SiO_2-Al_2O_3 ratio values is likely to reveal the fallacy at once, because the quotient is not affected by the mere addition or subtraction of a third component.

The author (3) has suggested a leaching factor β that consists of the ratio

$$\beta = \frac{ba_1 \text{ of leached horizon}}{ba_1 \text{ of parent material}}$$

In this instance, six analytical values are combined into one figure. The smaller β the greater is the relative leaching of K_2O and Na_2O with respect to Al_2O_3. If no relative loss of monovalent cations occurs, β equals unity. Data on the weathering of limestone may serve to illustrate the calculation of β (Table 3).

TABLE 3.—CALCULATION OF LEACHING VALUE β

Constituent	Rock		Soil	
	Per cent	Molecular value	Per cent	Molecular value
Al_2O_3	0.15	0.00147	7.59	0.0745
K_2O	0.20	0.00213	1.38	0.0147
Na_2O	0.06	0.00098	0.55	0.0089
ba_1	—	2.11	—	0.32
β	—	—	—	0.152

Owing to pronounced leaching of $CaCO_3$, the values for Al_2O_3, K_2O, and Na_2O are higher in the soil than in the original rock. Nevertheless, with respect to Al, the cations K and Na also have been drastically reduced.

Niggli (7) has suggested a system of symbols that is more comprehensive than the collection of ratios given above and yet not too cumbersome to handle. It differs from all other systems in that silica is treated independently.

Variability of Soil Types.—Every soil surveyor knows that the boundaries between soil types are not always so sharply defined as one might assume from an inspection of soil maps. This is due not so much to lack of accurate observation and mapping as it is the consequence of inherent variability of soil types (10).

On a priori grounds, any soil-forming factor that changes within a given geographic region causes soil variability, or, using another expression, reduces soil uniformity.

To express soil variability in quantitative terms, numerous graphic and mathematical devices are available. A simple graphical representation is known as the "soil line" or the "soil transect." On a chosen area, a line is drawn, along which, at regular intervals, the soil properties are measured and the results

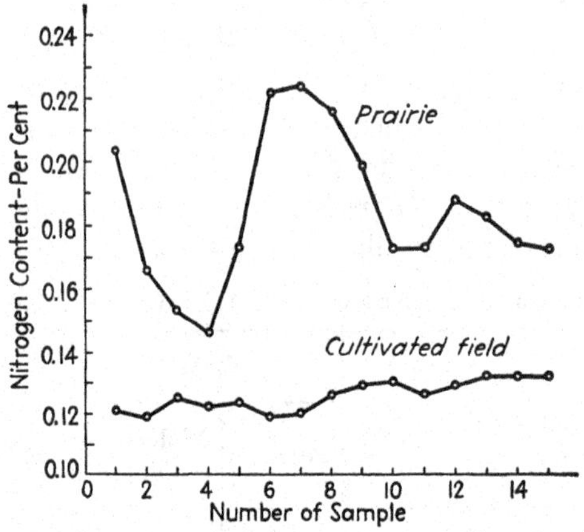

Fig. 13.—Soil transects for total nitrogen (0–7 inches depth). The upper curve represents a relatively heterogeneous soil, the lower curve a relatively homogeneous soil.

plotted as a function of distance. An actual case is illustrated in Fig. 13, which shows the variability of total nitrogen content (from 0 to 7 in. depth) in a virgin prairie and an adjoining cultivated field on the Putnam silt loam in Missouri. On the prairie, samples were collected at 120-ft. intervals; on the cultivated field, the samples were taken 30 ft. apart. The greater variability of the prairie series as compared with that of the cultivated field is clearly brought out.

If the samples are collected at random and in large numbers, quantitative characterization may be easily obtained with the aid of statistical formulas. The average nitrogen content of 73 samples of the aforementioned prairie amounts to 0.197 ± 0.0027 per cent. The corresponding value for the cultivated field is

0.129 ± 0.0013. The mean errors 0.0027 (prairie) and 0.0013 (cultivated field) represent numerical indexes of the variability within the two areas. In Fig. 14, the variability is depicted in the form of frequency diagrams. The smooth curves have been drawn according to the well-known Gaussian distribution equation

$$y = \frac{h}{\sqrt{\pi}} e^{-h^2x^2} \tag{6}$$

y denotes the relative frequency and x the relative class interval. The parameter h, corresponding to the maximum height of the

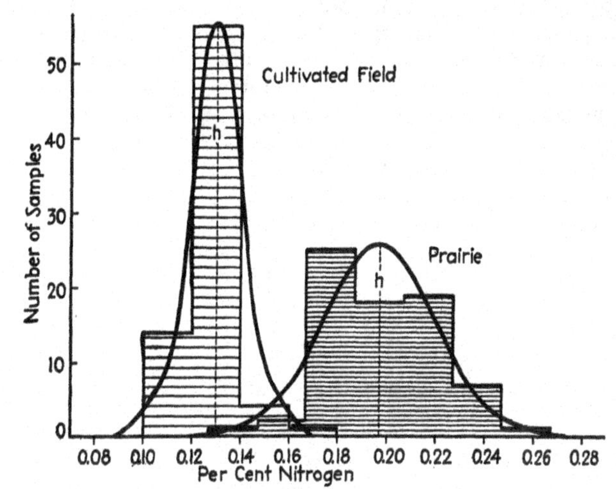

FIG. 14.—Illustration of soil variability by means of frequency curves. The higher the maximum of the curve the greater is the uniformity of the soil.

curves, may be used as a quantitative index of soil uniformity. The values of h are 0.610 for the prairie and 1.297 for the cultivated field.

Not all soil properties follow the Gaussian curve. For example, in the vicinity of limiting values of soil properties, unsymmetrical distribution curves of the Poisson type may be found. Curves that are extremely flat have low values of h and are suggestive of ill-defined soil types.

Literature Cited

1. GESSNER, H.: "Die Schlämmanalyse," Akademische Verlagsgesellschaft m. b. H., Leipzig, 1931.

2. HARRASSOWITZ, H.: Laterit, *Fortschr. Geolog. und Paleont.*, **4**:253–566, 1926.
3. JENNY, H.: Behavior of potassium and sodium during the process of soil formation, *Missouri Agr. Expt. Sta., Research Bull.* 162, 1931.
4. JENNY, H.: "Properties of Colloids," Stanford University Press, Stanford University, Calif., 1938.
5. MARBUT, C. F.: Soils of the United States, *Atlas of American Agriculture*, Part III, U. S. Government Printing Office, Washington, D. C., 1935.
6. MARSHALL, C. E.: Studies on the degree of dispersion of the clays, *J. Soc. Chem. Ind.*, **50**:444–450, 1931.
7. NIGGLI, P.: "Die Gesteinsmetamorphose," Verlagsbuchhandlung Gebrüder Borntraeger, Berlin, 1924.
8. ROBINSON, G. W.: "Soils: Their Origin, Constitution and Classification," Thomas Murby, London, 1932.
9. STEELE, J. G., and BRADFIELD, R.: The significance of size distribution in the clay fraction, *Am. Soil Survey Assoc., Bull.* **15**:88–93, 1934.
10. YOUDEN, W. J., and MEHLICH, A.: Selection of efficient methods for soil sampling, *Contrib. Boyce Thompson Inst.*, **9**:59–70, 1937.

CHAPTER III

TIME AS A SOIL-FORMING FACTOR

The estimation of relative age or degree of maturity of soils is universally based on horizon differentiation. In practice, it is generally maintained that the larger the number of horizons and the greater their thickness and intensity the more mature is the soil. However, it should be kept in mind that no one has ever witnessed the formation of a mature soil. In other words, our ideas about soil genesis as revealed by profile criteria are inferences. They are theories, not facts. This accounts for the great diversity of opinion as to the degree of maturity of specific soil profiles. It is well known that certain eminent pedologists take objection to the general belief that chernozems are mature soils; others consider brown forest soils and gray-brown-podsolic soils merely as immature podsols. The list of controversial soil types is quite long. Whatever the correct interpretation may be, it is evident that the issues center around the factor time in soil formation.

General Aspects of Time Functions.—If the fundamental equation of soil-forming factors

$$s = f(cl, o, r, p, t, \cdots) \qquad (4)$$

is evaluated for time, we obtain an expression of soil-time functions as follows:

$$s = f\ (\text{time})_{cl,o,r,p,\ldots} \qquad (7)$$

This equation states that the magnitude of any soil property (s type) is related to time. If we wish to ascertain accurately the nature of a time function, all remaining soil-forming factors must be kept constant. If they vary effectively, at one time or another, the trends of soil development are shifted, new processes are instigated, and we must start counting anew. The requirement of constancy of soil-forming factors is easily accomplished with controlled laboratory or field experiments. Under natural conditions, especially in the absence of historic records, we must be satisfied with approximate solutions of Eq. (7).

Studies on Experimental Weathering.—Hilger (8) exposed uniform rock particles of from 10 to 20 mm. diameter to atmos-

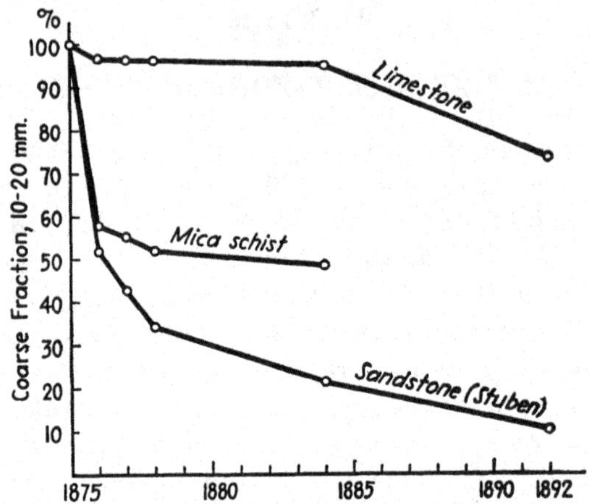

Fig. 15.—Hilger's experimental weathering series. Coarse particles of limestone are much more resistant than those of sandstone (var. Stuben).

pheric influences for a period of 17 years. The percentage of original particles left at various time intervals and the amounts of fine earth (particles less than 0.5 mm. diameter) formed are

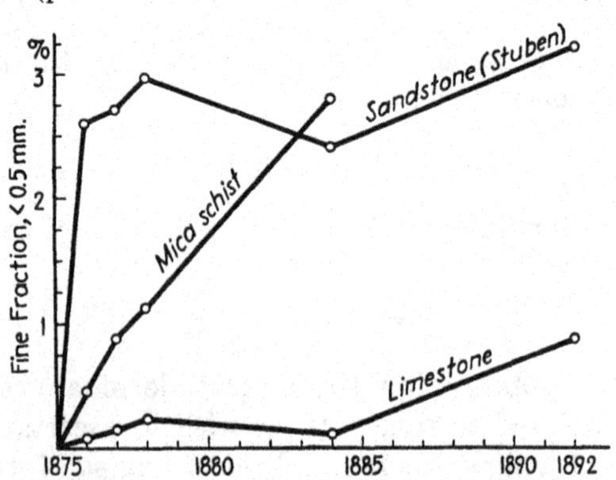

Fig. 16.—Hilger's experimental weathering series. This figure shows the amount of fine earth formed, expressed in per cent of original rock material.

shown in Figs. 15 and 16. The great variation in rates of physical weathering of different rock species is striking. Limestone was the most and sandstone the least resistant material.

Nearly 90 per cent of the coarse fraction of the sandstone disappeared within less than two decades. At the end of the 17-year period, Bissinger (4) determined the chemical composition of Hilger's weathering series. The results of both the original and the weathered rocks are given in Table 4 in terms of ratios. Differential weathering is clearly indicated. Limestone suffered greater relative losses of potassium and sodium than sandstone. The same relationship exists for the relative leaching of silica. In all cases, the fine earth has a lower *sa* value than the original rock, which indicates relative enrichment of aluminum. These changes are probably due in part to chemical leaching and in part to selective physical disintegration of the rock constituents. It is of interest to note that in these specific cases the sandstone exceeded the limestone in physical weathering but lagged behind in chemical weathering.

TABLE 4.—CHEMICAL DATA ON ROCK WEATHERING
(17-year period)

Type of material		$\frac{K_2O + Na_2O}{Al_2O_3} = ba_1$		$\frac{SiO_2}{Al_2O_3} = sa$	
		Absolute	Relative	Absolute	Relative
Limestone	Rock	5.00	1	17.08	1
	Fine earth	0.514	0.103	7.33	0.429
Mica schist	Rock	0.418	1	5.88	1
	Fine earth	0.288	0.690	4.87	0.829
Sandstone (Stuben)	Rock	0.247	1	40.1	1
	Fine earth	0.275	1.111	23.9	0.596

Further Estimates of Rates of Weathering.—Some 60 years ago, Geikie (5) made a survey of dated tombstones in Edinburgh churchyards. A marble stone with the inscription 1792 was partly crumbled into sand, whereas, in the case of a clay slate stone, the lettering was still sharp and showed scarcely any change. A sandstone of good quality indicated practically no effect of weathering during a period of 200 years. Goodchild (6), in 1890, published some notes on the weathering of limestones based on observation periods of 50 years or less. He calculated initial weathering rates as shown in Table 5.

TABLE 5.—WEATHERING RATES OF TOMBSTONES

Type of Tombstone	Number of Years Necessary to Produce 1 In. of Weathering
Kirkly Stephen	500
Tailbrig "macadam"	300
Penrith limestone	250
Askrigg limestone	240

Moisture is considered to be the major agent responsible for decay of building materials. Malécot as quoted by Merrill (14), emphasizes the durability of buildings and monuments in the dry climates of Egypt and Sicily; whereas the same materials transported to France or England soon reveal signs of decay.

TABLE 6.—AGE AND WEATHERING OF BUILDING MATERIALS

Type of building	Age, years	Notes on weathering conditions
Sandstone from Brochterbeck		
Church in Riesenbeck	About 100	In good condition
St. Catherine's Church in Osnabrück	About 550	Slight weathering
St. Mary's Church in Osnabrück	About 770	Strongly weathered in parts
Ruins of Castle Tecklenburg	About 900	Most of it is strongly weathered
Sandstone from Rothenburg		
Canal lock near Alt-Friesack	About 55	Significant traces of weathering
Same, near Spandau	About 80	Rather strongly weathered
Same, near Liepe	About 100	Very strongly weathered
Porphyry from Nahetal, Rhineland		
City hall in Kreuznach	150	No significant trace of weathering
High school in Kreuznach	400	Distinct surface weathering, no change in interior

Hirschwald (10) has made a systematic study of weathering processes on a large number of old buildings in central Europe. By limiting comparisons to materials from known quarries, he was able to establish semiquantitative relationships between time and degree of weathering (Table 6).

Soil Formation on the Kamenetz Fortress.—The Kamenetz fortress in Ukraine, U.S.S.R., was built in 1362 and remained in use until 1699, when its strategic position came to an end. The buildings were neglected, and the structure disintegrated. It may be assumed that, on the high walls and towers, weathering continued undisturbed throughout the subsequent centuries. In 1930, Akimtzev (2) investigated the soils formed on top of the

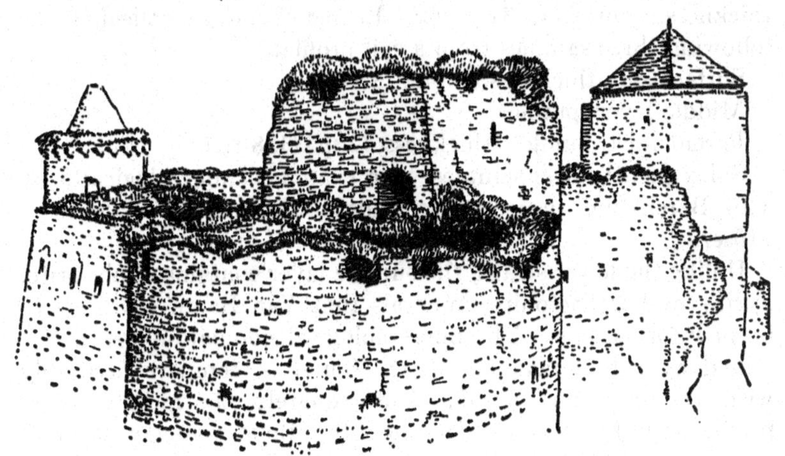

FIG. 17.—View of Kamenetz fortress from the west. (*After photograph*).

walls of the Dennaya tower (see Fig. 17) of the old fortress that had been constructed with calcareous slabs. He compared the weathered material with near-by soils derived from Silurian limestone (Table 7). Both soils are of the humus-carbonate type (rendzina), and their physical and chemical properties are remarkably alike. On the tower, soil development has been very

TABLE 7.—COMPARISON OF HISTORICAL AND NATURAL SOILS (KAMENETZ FORTRESS)

	Soil on fortress	Natural soil in vicinity of fortress
Depth of soil, cm	10–40	8
Humus, per cent	3.5	3.8
Clay, per cent	50–56	53
$CaCO_3$, per cent	5	About 5
Exchangeable Ca, per cent	0.85	0.89
pH	7.7	7.67

rapid, an average thickness of 12 in. of soil having been reached in 230 years.

Volcanic Soils.—On Aug. 26 and 27, 1883, occurred the stupendous volcanic eruption of Krakatao in the Sunda Strait, between Java and Sumatra (19). Enormous quantities of dust were projected into the atmosphere, covering the neighboring island, Lang-Eiland, with volcanic material over 100 ft. (30 m.) in thickness. On Oct. 31, 1928, Ecoma Verstege collected the following three samples from a soil profile:

Surface soil, thickness 35 cm. (13.8 in.),
Middle layer, pumice,
Parent rock, pumice, thickness 42 m. (138 ft.).

Subsequently, the samples were examined mineralogically by Van Baren and subjected to a detailed chemical analysis by Möser.

Unlike the lower strata, the surface layer contained anhydrite, pyrite, and wollastonite, minerals that Van Baren considers to be new formations. The microbiological population was determined by Schuitemaker and was found to compare favorably with that of a garden soil. Unfortunately, the morphological profile description is very meager, but there does not seem to have been any appreciable laterization. On the other hand, the chemical analyses (Table 8) clearly indicate tropical weathering. In spite of luxuriant vegetation, the nitrogen and organic-matter contents of the soil are low. During soil formation, the SiO_2-Al_2O_3 ratio has distinctly narrowed, indicating a preferential leaching of silica over alumina, which is supposedly a characteristic feature of tropical soil formation. The leaching factor β is 0.776, which compares favorably with that of podsolized soils of the humid temperate region. This is indeed a remarkable removal of potassium and sodium for the brief period of 45 years. The increase in fine particles and in moisture, especially in that driven off above 110°C., also speaks for significant chemical and physical conversions of rock into soil.

Hardy (7) has reported a significant accumulation of nitrogen and organic matter on recent volcanic-ash soils of the Soufrière district in St. Vincent, British West Indies. Fourteen years after the last volcanic eruption, the surface foot layer of soils at an altitude of 2,000 ft. contained 0.022 to 0.035 per cent nitrogen and 1.0 to 2.0 per cent organic matter. In 1933, 30 years after the eruption, the reforested region had attained comparative stability. Surface-soil samples collected at 10 different sites showed an average organic-matter content of 2.1 per cent and an average nitrogen content of 0.10 per cent (carbon-nitrogen ratio = 12.2) in the upper six-inch layer, values that are comparable with those for most of the cultivated soils of St. Vincent. "Thus," writes Hardy, "within 10 to 20 years, sterile volcanic ash may give rise to fertile soil under the prevailing circumstances."

TABLE 8.—CONDENSED DATA OF A LANG-EILAND SOIL, 45 YEARS OLD
(*Van Baren, et al.*)
Annual rainfall = 262 cm. (103 in.)
Annual temperature = 27.8°C. (82°F.)

Constituents	Rock	Middle layer	Surface soil
SiO_2, per cent	67.55	65.87	61.13
Al_2O_3, per cent	15.19	16.31	17.24
Fe_2O_3, per cent	1.52	1.74	2.56
FeO, per cent	2.15	2.05	2.59
CaO, per cent	2.89	3.07	3.61
Na_2O, per cent	4.47	4.01	3.90
K_2O, per cent	1.95	1.53	1.78
CO_2, per cent	—	—	0.04
H_2O, above 110°, per cent	2.46	3.17	3.25
H_2O, below 110°, per cent	0.04	0.33	1.53
Organic matter, per cent	—	—	0.45
Nitrogen, per cent	0.018	0.012	0.035
pH	5.3	5.8	6.0
Particles below 20μ, per cent	—	22.4	26.1
Color	White	White	Gray
SiO_2:Al_2O_3 = sa	7.56	6.86	6.03
β (see p. 27)	—	0.816	0.776

Soil-time Relationships on Recent Moraines.—Since the absolute movements of a number of alpine glaciers during the last hundred years are fairly accurately known, the study of their moraines provides good quantitative data on rates of soil formation. Figure 18 illustrates the relative positions of the Mittelberg Glacier in Tirol and two terminal moraines that were deposited in 1850 and 1890. Miss Schreckenthal (16), in 1935, studied a number of soils in this region (Table 9). In spite of seemingly unfavorable climatic conditions, particularly low temperatures, the moraines have been significantly altered within a period of 80 years. Soil acidity developed rapidly, silt became relatively abundant, and even some clay was formed. Notwithstanding the paucity of the flora, soil nitrogen is now high.

Hoffmann (16), working in the same vicinity, reports nitrogen analyses that are presented in graphic form in Fig. 19. The nitrogen-time curve appears to ascend in logarithmic manner, tending to approach a maximum. Although the data are quite

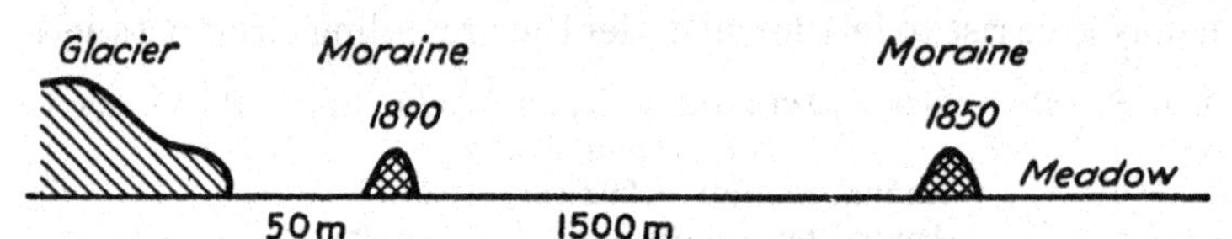

FIG. 18.—This graph shows in a sketchlike manner the front of the Mittelberg Glacier in Tirol and the position of two terminal moraines. Distances are given in meters.

scanty, they demonstrate, nevertheless, the rapid accumulation of soil nitrogen at high altitudes.

Time Functions Pertaining to Horizon Development.—In the section on Vegetation in Chap. VII we shall have occasion to consider in detail an eighty-year-old pine-hardwood succession in the Harvard Forest (Fig. 112, page 230). Under pine, the thickness of the forest floor gradually increased during the first 40 years and then remained approximately constant for the same

TABLE 9.—TIME SERIES OF MORAINIC SOILS (*Schreckenthal*)

Locality	Age, years	Depth of sampling, centimeters	pH	*N*, per cent	Composition of fine earth (<2 mm.), per cent	
					2–20μ	<2μ
Sand in front of glacier	0?	10	6.18	0.012	0.8	0.8
Skeleton soil above side moraine	—	—	6.65	0.009	—	—
Between glacier and 1890 moraine	30–40	10	6.08	0.041	7	3
Between 1890 moraine and 1850 moraine:						
Under spruce	50–60	10	5.75	0.12	20	4
Under grass	50–60	10	6.0	0.11	15	4
Meadow in front of 1850 moraine	80	10	5.82	0.26	—	—

length of time, possibly indicating that an equilibrium status had been attained. Under hardwood, the depth of the duff layer was reduced, first rapidly, then slowly. The formation of the dark-brown zone in the mineral matrix was surprisingly rapid and reached a magnitude of 10 in. within 20 years.

Podsolization.—In his monograph on soil studies in the region of coniferous forest in north Sweden, Tamm (18) has presented some valuable data on the velocity of podsolization. Following the drainage of Lake Ragunda in 1796, perceptible podsolization occurred within a period of about 100 years. In one locality, the sandy parent material contained 0.5 per cent calcium carbonate, which was leached out to a depth of 10 in. under pine-heath

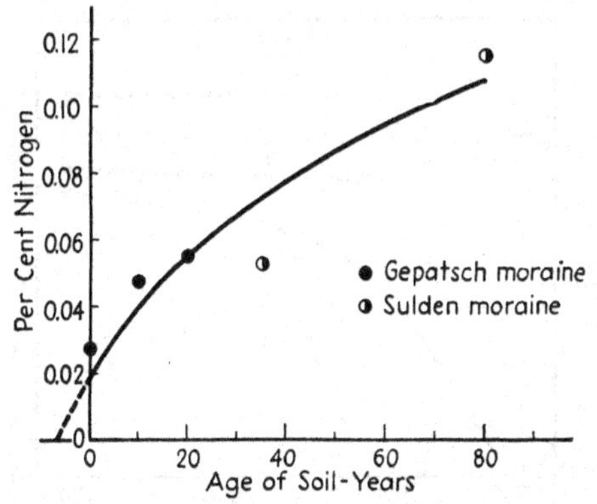

FIG. 19.—Nitrogen-time curve for morainic soils from the vicinity of the Mittelberg Glacier.

associations and to 25 in. under a mixed forest rich in mosses. In another locality, the sand deposit had become covered with a mattress of raw humus, and enough podsolization had taken place to permit a photographic recording of a thin bleached A_2 horizon and a dark orterde zone (*B* horizon). Tamm estimates that a normal podsol with 4 in. of raw humus, 4 in. of A_2 horizon, and from 10 to 20 in. of *B* horizon requires from 1,000 to 1,500 years to develop. Older soils, presumably 3,000 to even 6,000 or 7,000 years of age, do not exhibit horizons of greater magnitudes. Evidently in these localities profile formation has come to a standstill.

We owe to Tamm a series of chemical data on podsols from alluvial sand terraces the ages of which are known with considerable certainty. In Fig. 20, an attempt is made to present the analyses in the form of time graphs. The leaching factor β of the older profiles (β = 0.947) has values that are characteristic for podsols in general (Table 23, page 120). The behavior of the silica-alumina ratio (*sa*) of the soil also is instructive. Compared with the *C* horizon, the *sa* value of the bleached layer A_2 passes through a minimum at 100 years and then tends toward a maximum that corresponds to a relative accumulation of silica in the A_2 horizon. Tamm also determined the amounts of limonitic iron that may be taken as an index of the translocation of colloidal iron hydroxide. Bleaching signifies removal of $Fe(OH)_3$; darkening or reddening of the soil layers results from accumulation of this compound. To bring out the contrast more

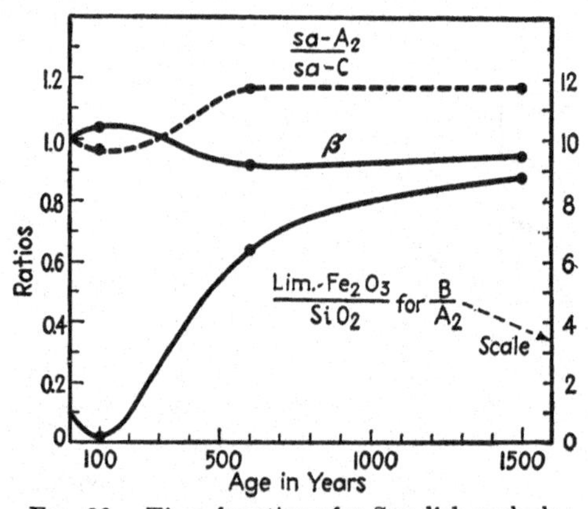

FIG. 20.—Time functions for Swedish podsols.

forcibly, the ratio limonitic Fe_2O_3/total SiO_2 of the *B* horizon was divided by the same ratio for the A_2 layer.

As may be seen from the lower solid curve in Fig. 20, there is in the initial phase of podsolization relatively more limonite in A_2 than in *B*, but at later stages the relationship is reversed. The magnitudes are indicative of substantial shifting of iron compounds.

A notable feature, common to all three curves, is the declining change of slope with increasing age. After drastic changes during the initial phases of profile formation, the soil characteristics are tending toward a more or less steady state, the equilibrium state or mature profile.

In this connection, two recent papers by Aaltonen (1) and by Mattson (13) are pertinent. Aaltonen studied the formation of the illuvial horizon in sandy soils of Finland. He concludes that it grows from the bottom up. In young soils, the colloidal particles are flocculated at greater depth than in old soils. Consequently, during the process of podsol formation, the portion of maximum colloid accumulation moves upward (Fig. 21). The behavior of the *A* horizon is not yet fully clarified. Aaltonen tentatively assumes that the thickness of the eluvial horizon increases with advancing age. Mattson fully supports Aaltonen's viewpoint. A schematic graph, adapted from these authors and portraying four profiles of varying age is reproduced in Fig. 21. Black shading denotes precipitation (*B* horizon), and cross-

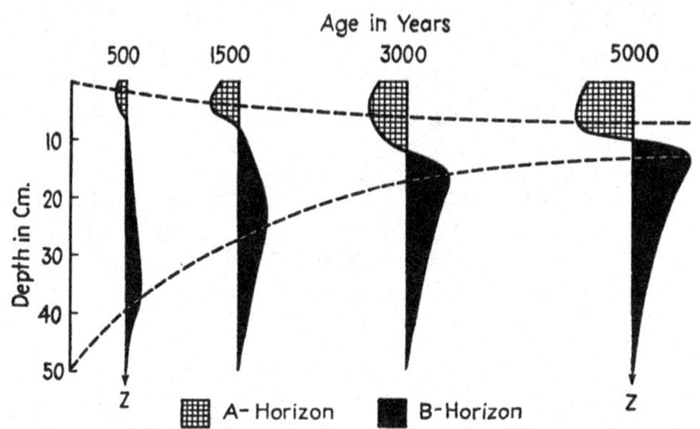

FIG. 21.—Formation of *A* and *B* horizons of a podsol profile as a function of age, according to Aaltonen and Mattson.

hatched areas indicate zones of removal (*A* horizon) of colloidal material. The distances vertical to the *Z*-axis correspond to amounts of colloidal sesquioxides. Mattson emphasizes the asymptotic trend of the horizon formation toward an equilibrium state.

Salisbury's Dune Series.—One of the most reliable time functions has been obtained by Salisbury (15) for the Southport dune system in England. The ages of the ridges have been assessed partly by examination of old maps (1610, 1736) and partly by descriptions. Moreover, the early arrival of *Salix repens* in the hollows between the dune ridges made it possible to utilize the number of annual rings as a check on the estimates. Salisbury's determinations of $CaCO_3$, pH, and organic matter are graphically summarized in Figs. 22 and 23.

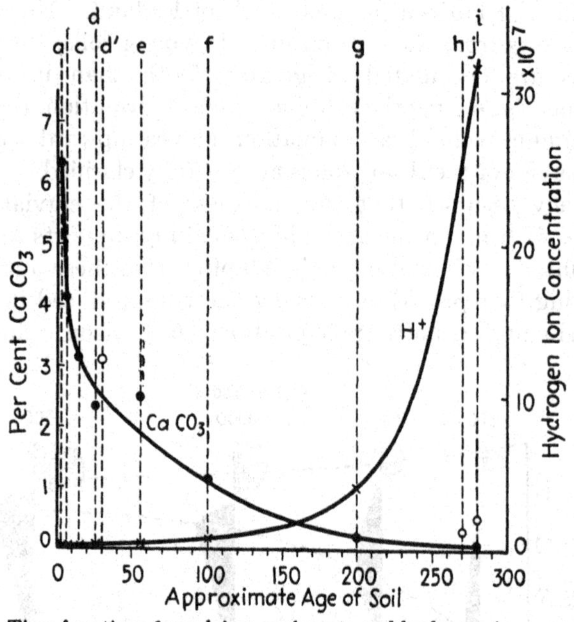

FIG. 22.—Time functions for calcium carbonate and hydrogen ion concentrations for Salisbury's dune series.

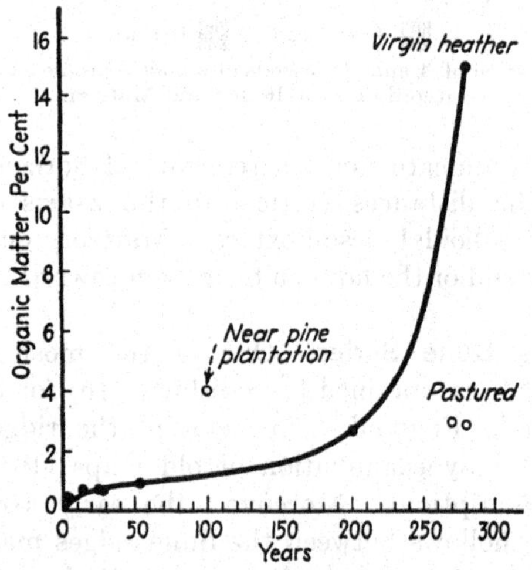

FIG. 23.—Time function for organic matter for Salisbury's dune series.

The *carbonate curve*, being of an exponential nature, is very steep at the beginning and becomes flatter as the age of the dune progresses. The *hydrogen ion concentration* of the initially calcareous dunes is necessarily low, and it remains so for over a century, in spite of very rapid removal of lime. Subsequent to the 200-year period, it rises very rapidly.

Salisbury also determined the organic-matter content of the dunes, or, more precisely, the loss on ignition, corrected for the carbonates. The curve shown in Fig. 23 runs somewhat parallel to the acidity curve. The organic-matter contents of some of the dunes appear erratic, on account of human interference, as

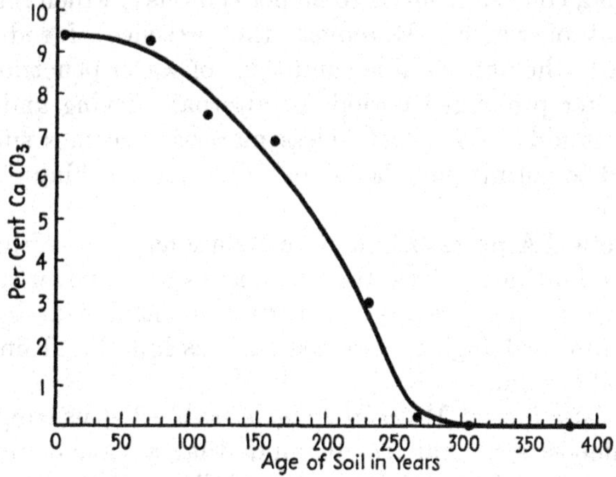

FIG. 24.—Leaching of calcium carbonate as a function of time in Dutch polders. (*Hissink.*)

Salisbury contends. The same dunes also show deviations in the calcium carbonate and acidity curves.

Leaching of Dutch Polders.—For centuries, the people of Holland have enlarged their agricultural area along the seacoast by building dams or dikes that prevent flooding of land during high tides. The muddy deposits thus wrested from the sea mark the initial phase of a process of soil development during which the salty and unproductive muds are transformed into fertile lands. In recent years, Hissink (11) has given an interesting account of the changes in soil properties that have taken place in the course of centuries. In Fig. 24 are plotted the percentages of calcium carbonate of the surface soil (from 0 to 8 or 10 in.) as a function of time. Originally the soil material contained from 9 to 10 per cent calcium carbonate, which completely disappeared from the surface soil within 300 years.

It is of interest to reflect on the shape of the curve, especially in comparison with the corresponding data of Salisbury. On the sand dunes, the highest rate of removal of calcium carbonate occurs during the initial years of soil development, whereas on the polders it is postponed for a century or two at least. The explanation of this profound difference must be sought in nature of the soil material. Salisbury's dunes are of sandy texture. Rain water readily percolates through the porous material and immediately produces leaching effects. The polders, on the other hand, contain from 60 to 80 per cent clay, which retards the movement of water. Moreover, the presence of sodium ions accentuates the unfavorable conditions of water penetration. It is only after prolonged periods of alternate drying and wetting that the muddy and structureless material becomes sufficiently organized to permit percolation of water and leaching of calcium carbonate.

Theoretical Aspects of Soil-time Relationships.—Although the data on quantitative time functions are scanty, nevertheless the knowledge at hand lends itself to a fruitful examination of fundamental pedological concepts such as initial and final states of the soil system.

When Does Parent Material Become Soil?—Let us suppose that a river deposits several feet of mud, that a violent dust storm covers a region with a thick blanket of silt, or that a volcano lays down a bed of ash. Are these deposits parent material or soil? Opinions are divided. A few soil scientists consider them soils, but the majority look upon them as parent material. The question immediately arises: If these deposits are not soils now, when do they become soils?

Most of us will agree that 1,000 years hence, the aforementioned deposits will have developed into soils. They will, perhaps, not be mature but in all probability will have acquired profile features that will be distinct enough to be seen in the field. In contrast to field examination, chemical analysis is able to detect eluvial and illuvial soil layers at a much earlier date. Clearly, the more refined our methods of observation, the sooner we shall be able to ascertain the change from geological material to soil. At what time, then, does parent material become soil?

The problem becomes more tangible on examination of specific soil-time functions such as those provided by Salisbury's dune series. The freshly formed dune unquestionably deserves the attribute parent material. The 280-year-old dune is distinctly a soil, because it has a *solum* and a *C* horizon. Focusing attention upon a specific soil characteristic, *e.g.*, the carbonate content, we see from the curve in Fig. 22 that it diminishes continuously as time advances. There is no break in the curve or a conspicuous point that might suggest the start of soil genesis. At the soil-formation time "zero," soil and parent material merge into each other. Mathematically speaking, we should say that parent material becomes soil after an infinitely small time interval dt. This idea leads to a simple and precise definition of parent material: *It is merely the state of the soil system at the soil-formation time zero.* It is for this very reason that in Chap. I we have defined parent material as the initial state of the soil system.

Time of soil formation is not necessarily identical with the "age of the country" or the "geological age of the land," as maintained in some publications. As soon as a rock, consolidated or unconsolidated, is brought into a new environment and acted upon by water, temperature, and organisms, it ceases to be parent material and becomes soil. Returning to the questions asked at the beginning of this section, we are forced to conclude that young riverbanks, fresh loess mantles, etc., are soils, unless they are being deposited under the very eyes of the observer.

Soil Maturity and the Concept of Soil Equilibrium.—Marbut (12) defines a mature soil as one "whose profile features are well developed." This definition is strictly morphological and may be applied directly in the field. Among students of soils, one frequently encounters a second definition that is enjoying increasing popularity. It rests on dynamic rather than morphological criteria and may be expressed as follows: "mature soils are in equilibrium with the environment." In this case, emphasis is not placed on profile descriptions but on soil-forming processes. More specifically, the concept has some bearing on the factor time in soil development. It alludes to the familiar equilibrium idea employed by chemists and physicists. It will prove profitable to analyze the equilibrium concept of soil maturity in the light of time functions.

In the first place, it should be kept in mind that not all soil components approach "maturity" at the same rate. In practice only a few soil characteristics are taken into consideration when questions pertaining to soil maturity are to be decided upon. They are, in the main, soil reaction, organic matter, lime horizon, clay accumulations, ortstein, or, more generally, the magnitudes of illuvial and eluvial horizons. Obviously, we may be dealing with partial equilibria only; the rest of the soil mass may still undergo further changes. In Hissink's carbonate curve of Dutch polders, the value of calcium carbonate becomes zero at $t = 300$ years; yet the remaining soil properties continue to vary with time.

In the second place, we need reliable equilibrium criteria. How are we going to decide whether or not equilibrium has been reached? In a purely formalistic manner, we may postulate that a soil, or, more specifically, a soil characteristic s, is in equilibrium with the environment when its indicatrix no longer changes with time. The component of the system is then at rest or stabilized. In mathematical parlance, one would write

$$\frac{\Delta s}{\Delta t} = 0$$

where Δs denotes a change in a soil property and Δt a time interval. This definition is restricted to the soil as a macroscopic system. It disregards the behavior of individual atoms and molecules such as thermal agitation and Brownian movement, which never cease.

Applying the above criterion to soil-property-time functions, equilibrium would be reached when the curves become and remain flat, an indication that the rate of change is zero (Fig. 25*a*). It does not necessarily follow that the soil characteristic is at an absolute standstill, for, if we choose Δt sufficiently large, daily or seasonal fluctuations would not affect the average slope of the curve (Fig. 25*b*). Tamm's curves, and Hissink's calcium carbonate curve might be cited as examples of functions that reveal equilibrium in regard to certain soil characteristics.

A word might be said about the choice of time scales. For absolute evaluation, the year, or any multiple of it, appears to be the appropriate unit. Since certain soils reach maturity much more quickly than others, it may become desirable, for purposes of comparison, to use relative time units, *e.g.*, fractions of maturity ages (Fig. 26). Instead of speaking of young and old soils, one would use terms like "immature," "mature," or "degrees of maturity" ("one-half mature," etc.).

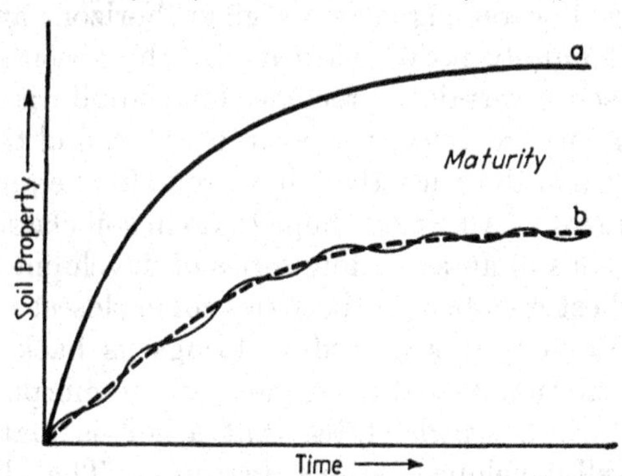

FIG. 25.—Schematic illustration of the variation of a soil property with time. The level branches of the curves signify maturity. The lower curve (*b*) indicates a soil property that is subjected to periodic fluctuations, such as seasonal variations.

Satisfactory as the formalistic definition of equilibrium is in principle, it defies observational verification for the majority of soils because of their slow speed of reaction. Unless we have accurate records that extend over centuries and millenniums, we are never sure whether we are dealing with true equilibria or merely slow reaction rates.

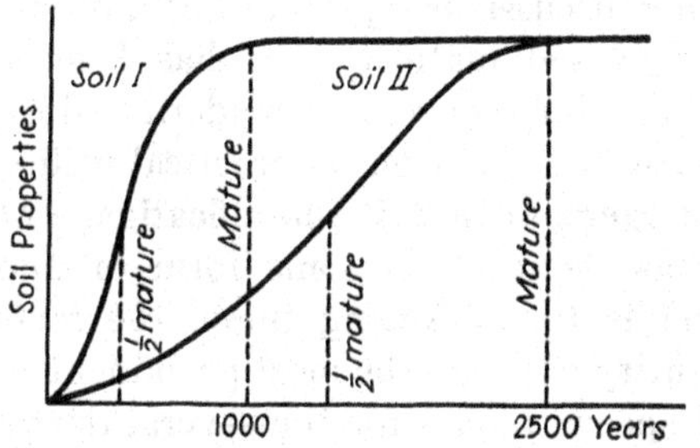

FIG. 26.—Soil property-time functions with reference to degrees of maturity.

There is, however, another avenue of approach. The most active agency in soil-profile formation is percolating water. As long as water passes through the *solum*, substances are dissolved, translocated, precipitated, and flocculated, and the soil is not in a state of rest. This "force," the penetrating rain water, is counteracted by evaporation, transpiration of vegetation, and by impenetrable horizons such as clay pans and hardpans, which are themselves the result of soil formation.

Now, we can learn with a reasonable degree of assurance whether or not water percolates a given horizon; and, although we cannot definitely prove maturity by the absence of percolation, its presence certainly precludes final equilibria. Most soils of humid regions are not impervious to water, and therefore they are not mature in the sense that they are in final equilibrium with the environment. All we can hope to do in soil classification is to arrange the soils in an ascending series of development, implying that the highest members in the series come closest to maturity.

In point of fact, this procedure brings us back to the first-mentioned criterion of maturity, namely, the morphological definition of Marbut, which states that a soil is mature when it possesses well-developed profile features. The dynamic and the morphologic criteria of soil maturity are in fair harmony as long as soils of equal constellations of *cl*, *o*, *r*, and *p* are being considered.

Uncertainty arises in comparisons of soils of different origins or different climatic regions, for the morphologic and the dynamic criteria may stand in contradistinction to each other. Theoretically, at least, soils may be in equilibrium with the environment without having marked profile features, and, vice versa, soils may have well-developed profiles without being in equilibrium.

The foregoing discussions represent but a fragmentary analysis of the nature of soil maturity. It has been the purpose to elucidate this cardinal concept of modern pedology rather than to pass dogmatic judgment on its practical utility.

Time as an Element in Soil Classification.—Most systems of soil classification contain, in some form or another, the idea of soil-forming factors. Among these, the factor time or the degree of maturity occupies the most prominent role. Shaw (17) has proposed special names for the several degrees of maturity:

Solum crudum (raw soil),

Solum semicrudum (young soil, only slightly weathered),

Solum immaturum (immature soil, only moderately weathered),

Solum semimaturum (semimature, already considerably weathered),

Solum maturum (mature soil, fully weathered).

A consistent classification of soils according to degrees of maturity is obtained for conditions of constancy of climate, organisms, parent material, and topography. In accordance with Eq. (7) the soil profile then becomes solely a function of time. Shaw's San Joaquin family of soils in California provides a good illustration. This soil family has developed on broad alluvial fans composed of granitic rock debris. The climate is

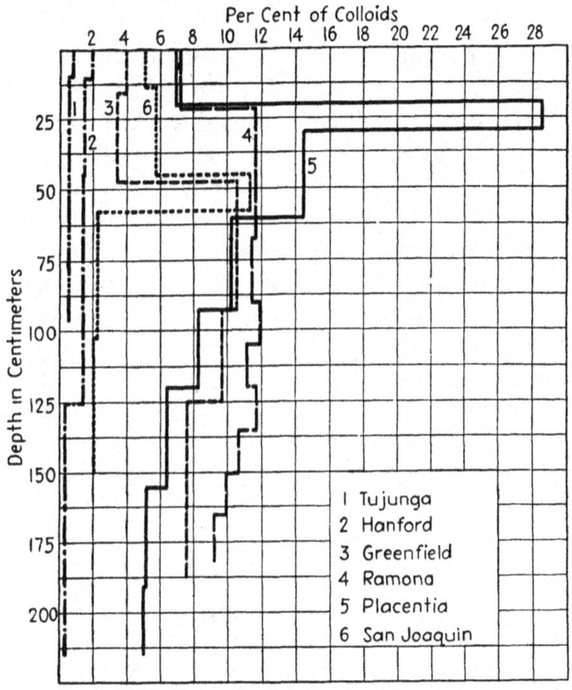

Fig. 27.—Percentage of inorganic colloidal material of the San Joaquin family as a function of depth.

semiarid with a rainfall ranging between 10 and 20 in. The native cover is shrubs, herbs, and grass with occasional oak trees. Shaw divides the sequence of development states into six phases

Tujunga sand → Hanford fine sand → Greenfield sand → Ramona sandy loam → Placentia sandy loam → San Joaquin sandy loam

Tujunga is the youngest member, with the least developed soil profile; San Joaquin represents the climax, the mature soil. Soil reaction varies little throughout the soil-forming process; it fluctuates between pH 7.2 and 8.2. The most outstanding feature is the colloid content, particularly its distribution among the various horizons of the profile (Fig. 27). In the *A* horizon, the colloid content increases from 0.83 per cent in the Tujunga to 7 per cent in the older members. The most drastic changes occur in the *B* horizon where, in the advanced stages, the clay accumulation gives rise to a reddish, indurated, impervious hardpan.

The factor time or age also figures prominently in the current system of classification of soils of the United States, which is an offspring of the Russian school of thought. Kellogg (3) and his associates classify all soils into three great orders: zonal soils, intrazonal soils, and azonal soils. The last two orders contain, among other soil types, those soils that, owing to their youth, have not fully developed profiles. Marbut's system of soil classification also includes the time factor. His differentiation of soils into pedocals and pedalfers (see page 190) applies only to soils that have mature profiles.

Notwithstanding the great popularity of systems of classification based on maturity series, we should not lose sight of their speculative and hypothetical nature. At present we have no reliable method of determining accurately the degree of maturity of any soil type.

Literature Cited

1. Aaltonen, V. T.: Zur Stratigraphie des Podsolprofils, II, *Commun. Instituti Forestalis Fenniae*, **27.4**:1–133, Helsinki, 1939.
2. Akimtzev, V. V.: Historical Soils of the Kamenetz-Podolsk Fortress, *Proc. Second Intern. Congr. Soil Sci.*, **5**:132–140, 1932.
3. Baldwin, M., Kellogg, C. E., and Thorp, J.: Soil Classification, *Yearbook of Agriculture*, 1938: 979–1001, U. S. Government Printing Office, Washington, D. C.
4. Bissinger, L.: Über Verwitterungs-Vorgänge, Dissertation, Erlangen, 1894.
5. Geikie: Rockweathering, as illustrated in Edinburgh churchyards, *Proc. Roy. Soc. Edinburgh*, **10**: 518–532, 1880.
6. Goodchild, J. G.: Notes on some observed rates of weathering of limestones, *Geol. Mag.*, **27**: 463–466, 1890.
7. Hardy, F.: Soil erosion in St. Vincent, B.W.I., *Trop. Agr.* (*Trinidad*), **16**: 58–65, 1939.
8. Hilger, A.: Über Verwitterungsvorgänge bei krystallinischen und Sedimentärgesteinen, *Landw. Jahrb.*, **8**:1–11, 1897.

9. Hilger, A., and Schütze, R.: Über Verwitterungsvorgänge bei krystallinischen und Sedimentärgesteinen, II, *Landw. Jahrb.*, **15**:432–451, 1886.
10 Hirschwald, J.: "Die Prüfung der natürlichen Bausteine auf ihre Wetterbeständigkeit," Berlin, 1908.
11. Hissink, D. J.: The reclamation of the Dutch saline soils (Solonchak) and their further weathering under humid climatic conditions of Holland, *Soil Sci.*, **45**: 83–94, 1938.
12. Marbut, C. F.: Soils of the United States, *Atlas of American Agriculture*, Part III, U. S. Government Printing Office, Washington, D. C., 1935.
13. Mattson, S., and Lönnemark, H.: The pedography of hydrologic podsol series I, *Ann. Agr. Coll. Sweden*, **7**:185–227, 1939.
14. Merrill, G. P.: "Stones for Building and Decoration," New York, 1903.
15. Salisbury, E. J.: Note on the edaphic succession in some dune soils with special reference to the time factor, *J. Ecol.*, **13**: 322–328, 1925.
16. Schreckenthal-Schimitschek, G.: Der Einfluss des Bodens auf die Vegetation im Möranengelände des Mittelbergferners (Pitztal, Tirol), *Z. Gletscherkunde*, **23**: 57–66, 1935.
17. Shaw, C. F.: Profile development and the relationship of soils in California, *Proc. First Intern. Congr. Soil Sci.*, **4**:291–397, 1928.
18. Tamm, O.: Bodenstudien in der Nordschwedischen Nadelwaldregion, *Medd. Statens Skogsförsöksanstalt*, **17**:49–300, 1920.
19. Van Baren, J.: Properties and constitution of a volcanic soil, built in 50 years in the East-Indian Archipelago, *Comm. Geol. Inst. Agr. Univ. Wageningen, Holland*, 17, 1931.

CHAPTER IV

PARENT MATERIAL AS A SOIL-FORMING FACTOR

The lower strata of a soil profile, especially those immediately below the *B* horizon, are commonly designated as *C* horizons. In many soils of northern regions, such as podsols, the selection of the *C* horizon offers no difficulties, because the *A* and *B* horizons are sharply differentiated from the unweathered material. In temperate regions, the selection of the *C* horizon is often ambiguous, especially if the soils are derived from sedimentary deposits. In practice, the difficulties are circumvented by distinguishing several *C* horizons that are designated as C_1, C_2, C_3, or D_1, E_1, etc. In the tropics, where rock decomposition may occur to great depth, the problem of the proper choice of the *C* horizon is often so complex that doubt is cast on the universal applicability of the simple *ABC* scheme of soil-profile description.

A. DEFINITION AND METHOD OF APPROACH

Definition of Parent Material.—Most pedologists extend the purely descriptive definition of the *C* horizon to include the concept of parent material. They define the *C* horizon as parent material. In selecting lower strata of a soil profile as parent material, the implicit assumption is made that the *A* and *B* horizons were derived from material that is identical with that of the *C* horizon. Numerous cases are known where such an approach would lead to erroneous results. If a shallow layer of loess is deposited on diorite rock, *A* and *B* horizons may readily form in the loessial material, whereas the igneous substratum may remain practically unaltered. It would be fallacious to designate the diorite as *C* horizon. Or, let us suppose that a chernozem soil is transformed into a podsol as a consequence of drastic climatic changes. Should we give the attribute parent material to the *C* horizon of the podsol or to the original chernozem? In view of these logical and practical difficulties, we prefer to define parent material as the *initial state of the soil* *system* and thus avoid special reference to the strata below the soil, which may or may not be parent material. The selection of the initial state of the soil system follows most logically from considerations of time functions (see page 45).

In the introductory chapter, soil was defined in terms of its properties as follows:

$$F(cl', o', r', s_1, s_2, s_3, \cdots) = 0 \tag{2}$$

Accordingly, parent material, herein defined as the initial state of the soil system, and designated as π, may be mathematically expressed as follows:

$$F(cl'^0, o'^0, r'^0, s_1^0, s_2^0, s_3^0, \cdots) = 0 \tag{8}$$

The two equations are identical except that the affix zero indicates that in the case of parent material we are dealing with the initial value (at zero time of the soil formation) of each soil property.

As Overstreet* points out, Eq. (8) may at times lead into difficulties, because, logically, it would not permit comparative studies of rock weathering in widely different climatic regions. We might wish to investigate the decomposition of two granites of identical properties, one placed in the arctic region, the other near the equator. In those localities, the two rocks are, strictly speaking, not identical, because they differ, among other things, in temperature, which affects some of their properties, such as the refractive indexes of the mineral components. To overcome this shortcoming, we shall introduce an extension of Eq. (8) by stating that parent material is the initial state of the soil system referred to some arbitrary standard state (*e.g.*, normal temperature and pressure). Thus, the two granites mentioned above become identical if referred to the arbitrarily chosen temperature of 18°C. Parent material so defined will be designated as p. In practice, the distinction between π and p is rather subtle and, in most cases, may therefore be neglected.

Determination of Parent Material.—It is only under special circumstances that the lower strata of a soil profile permit an exact quantitative evaluation of parent material as defined in the previous section. The exact composition of the initial

* Private communication.

state of a soil may be determined if the history of the soil, or, more precisely, the time functions of its properties are known. Such studies are restricted to soil families of known age. Another avenue of approach is given by soil-climate relationships, particularly soil-moisture functions. An illustration will be presented on page 121 in which the lime-rainfall curves of Alway are extrapolated to zero rainfall. The resulting value of 17.2 per cent CaO represents the calcium content of the parent material. Naturally, the reliability of such a determination depends entirely on the degree of accuracy of specific soil-property-climate functions. Generally speaking, the exact evaluation of the composition of the parent material involves considerable speculation and is the source of much uncertainty in the elucidation of soil-forming processes.

Weathering and Soil Formation.—Some of the leading pedologists of today lay great stress on the distinction between weathering processes and soil-forming processes. The former are said to be geologic; the latter are pedologic. Weathering includes solution, hydrolysis, carbonization, oxidation, reduction, and clay formation. Among the soil-forming processes, the following are listed: calcification, podsolization, laterization, salinization, desalinization, alkalization and dealkalization, formation of peat and poorly drained soils, including gleization (33). Since these are merely special types of chemical processes, their separation into geologic and pedologic groups appears neither convincing nor fruitful. The entire issue might be dismissed as being of purely academic interest, were it not for the fact that a practical consequence is involved, namely, the determination of parent material.

Pedologists who distinguish between geologic and pedologic processes do not regard granite, basalt, limestone, and consolidated rocks in general as parent materials. They maintain that only the weathered portions furnish material for soil-building purposes, and only these deserve the name "parent material." To mention a case in point, it has been contended that the relationship between clay content of soil and annual temperature, to be discussed on page 150, refers to a geologic and not to a pedologic problem, because it deals with the formation of parent material rather than of soil. Whatever the merits of this new approach may be, it frustrates the climatic theory of soil formation. Since weathering is controlled by moisture and temperature, it follows that the formation of parent material also becomes a function of climate. Parent material could no longer be treated as an independent variable and therefore would cease to be a soil-forming factor. All soil property-climate functions to be discussed in the following chapters become meaningless, because they would have to be considered as the result of complex combinations of soil-forming and parent-material-forming processes.

A most serious difficulty for the drawing of sharp distinctions between weathering reactions and soil-forming processes is presented by the continuation of weathering during soil development. Not only soil but also parent material would become a function of time. We could no longer speak of soil-maturity series (*e.g.*, Shaw's family, Bray's sequences), because such a concept presupposes constancy of parent material. It is true, of course, that for a given development series such as

$$\text{Rock} \rightarrow \text{Weathered rock} \rightarrow \text{Immature soil} \rightarrow \text{Mature soil}$$

our definition of parent material as the initial state of the soil system permits the choice of either rock or weathered rock as the starting point. But, once the initial state has been chosen, it must be treated as a constant and not as a variable. Such is possible only if climatically controlled weathering reactions are included among the soil-forming processes.

Should an arbitrary differentiation between geologic and pedologic processes be insisted upon, the functional definition of soil, given in Chap. I, would readily lend itself to such purpose. We may formulate: *Any reaction taking place in soils that is functionally related to soil-forming factors* [Eq. (4)] *is a soil-forming process.* Accordingly, natural phenomena such as volcanic eruptions, depositions of loess, and sedimentation in lakes and rivers are not soil-forming processes. They build up parent material and are classified as geologic phenomena. Likewise, any chemical reaction occurring in rocks that is not conditioned by Eq. (4) is arbitrarily excluded from the field of pedology.

Difficulties Encountered in Measuring Soil-Parent Material Functions.—For the study of the factor parent material the following equation is used

$$s = f(p)_{cl,o,r,t,\ldots} \qquad (9)$$

The greatest obstacle in evaluating quantitatively Eq. (9) lies in the task of assigning numerical values to different types of parent materials. We have no way of recording the properties of a diorite or of a sandstone in a single comparable figure. Such a number would have to embrace chemical composition, mineralogical constitution, texture, and structure of a rock. All we hope for in functional analysis of parent material is a correlation between soil properties and specific rock properties such as lime content, permeability, etc.

A notable attempt in this direction has been made by Prescott and Hosking (23), who correlated the mean clay content (from 0 to 27 in. depth) of red basaltic soils from eastern Australia with the mineralogical composition of the parent basalt. Rocks that were composed of from 51 to 58 per cent feldspars (orthoclase, albite, and anorthite) yielded soils containing from 54 to 55 per cent clay. If the total feldspar percentage rose to from 62 to 68 per cent, the content of clay amounted to from 63 to 76 per cent. The true nature of Prescott and Hosking's correlation is somewhat masked by variations in rainfall values.

In his report "Soils of Iowa," Brown (2) has established quantitative relationships between soil nitrogen, soil organic matter, and parent material. The data presented in Table 10 pertain to the Carrington series, a group of soils derived from glacial till, which to some extent has been modified by the presence of loess. Variations in the composition of this parent material are expressed by differences in soil texture. For surface soils, we may write the following equation:

$$\text{Nitrogen} = f\,(\text{texture})_{cl,o,r,t,\ldots} \qquad (10)$$

The constant soil-forming factors *cl*, *o*, *r*, and *t* are specified as follows:

cl = climate of Iowa (approximately identical for entire series),
o = prairie (now cultivated),
r = gently undulating to rolling,
t = unknown, but presumably the same for the entire series.

In the foregoing equation it is assumed that the texture of the surface soil defines the parent material of the surface soil. Brown's comparisons are summarized in Table 10. Both nitrogen and carbon increase as the soils assume a heavier texture. The mean values for *N* and *C* are significant inasmuch as the variability within soil types is considerably less than between textural groups.

Types of Parent Materials.—In Figs. 28, 29, and 30 are shown the distribution of various types of parent materials within the United States. These maps were constructed from Marbut's "Distribution of parent materials of soils" in his Soils of the

TABLE 10.—THE AVERAGE NITROGEN AND CARBON CONTENT OF SOIL TYPES OF DIFFERENT CLASSES OF THE CARRINGTON SERIES
(Surface soils, 0 to 6⅔ in. depth)

Texture of surface soils	Nitrogen, per cent	Organic carbon, per cent	C/N
Sand	0.028	0.40	14.1
Fine sand	0.043	0.58	14.5
Sandy loam	0.100	1.25	12.5
Fine sandy loam	0.107	1.32	12.5
Loam	0.188	2.21	12.2
Silt loam	0.230	2.68	11.7

United States in the *Atlas of American Agriculture*, Part III. According to Marbut,

> No attempt has been made to make it accurate in detail. In considerable areas, there may be some legitimate difference of opinion as to the source and character of the materials, such, for example, as on the plains of southern Idaho and parts of central Oregon and Washington. In central Texas, the western part of the area of residual accumulations from sandstones and shales contain areas of Great Plains materials and sands. The distribution of loess has been extended over areas about which there is no universal agreement. Notwithstanding these and many other areas of detail about which there is no universal agreement, the maps represent a mass of useful information.

B. SOIL FORMATION ON IGNEOUS ROCKS

Before the advent of the climatic theories of soil formation, parent material was considered the major soil-forming factor. In spite of a vast amount of work on the relationships between soil properties and underlying geological strata, relatively little information is at hand that can be interpreted in the light of functional analysis. As a rule, the correlations are suffering from lack of control of soil-forming factors, particularly climate.

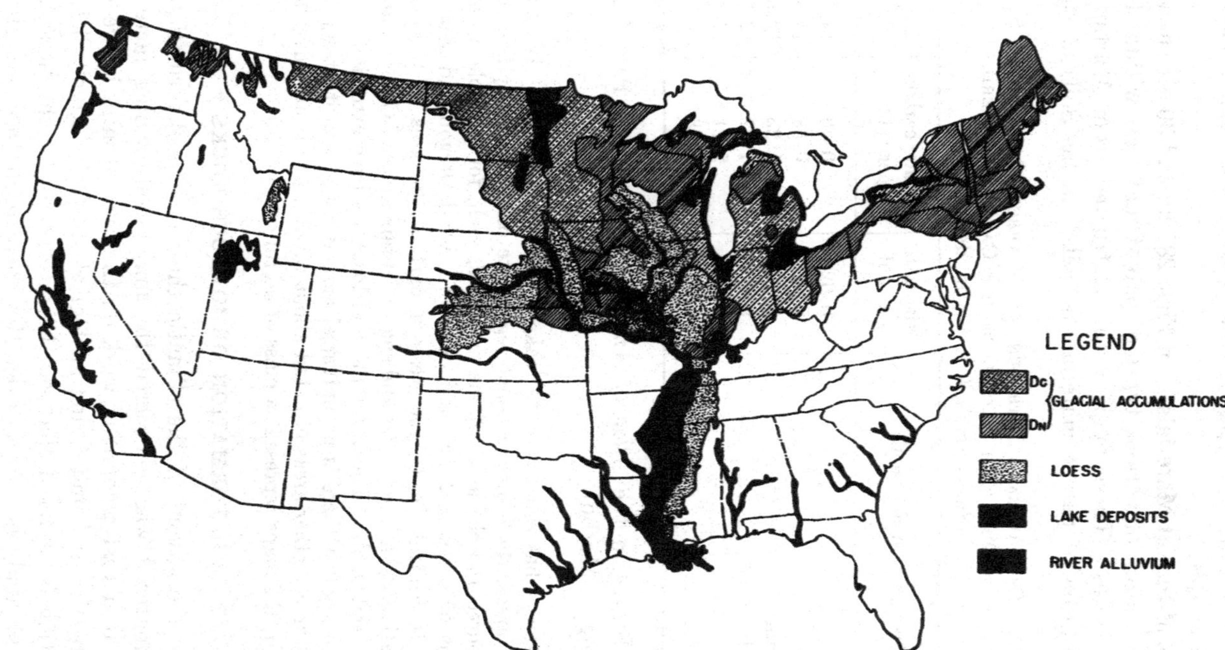

FIG. 28.—Distribution of parent materials of soils. Areas of unconsolidated rocks predominantly of Pleistocene origin. *Dc* = highly calcareous glacial accumulations; *Dn* = slightly or noncalcareous glacial accumulations.

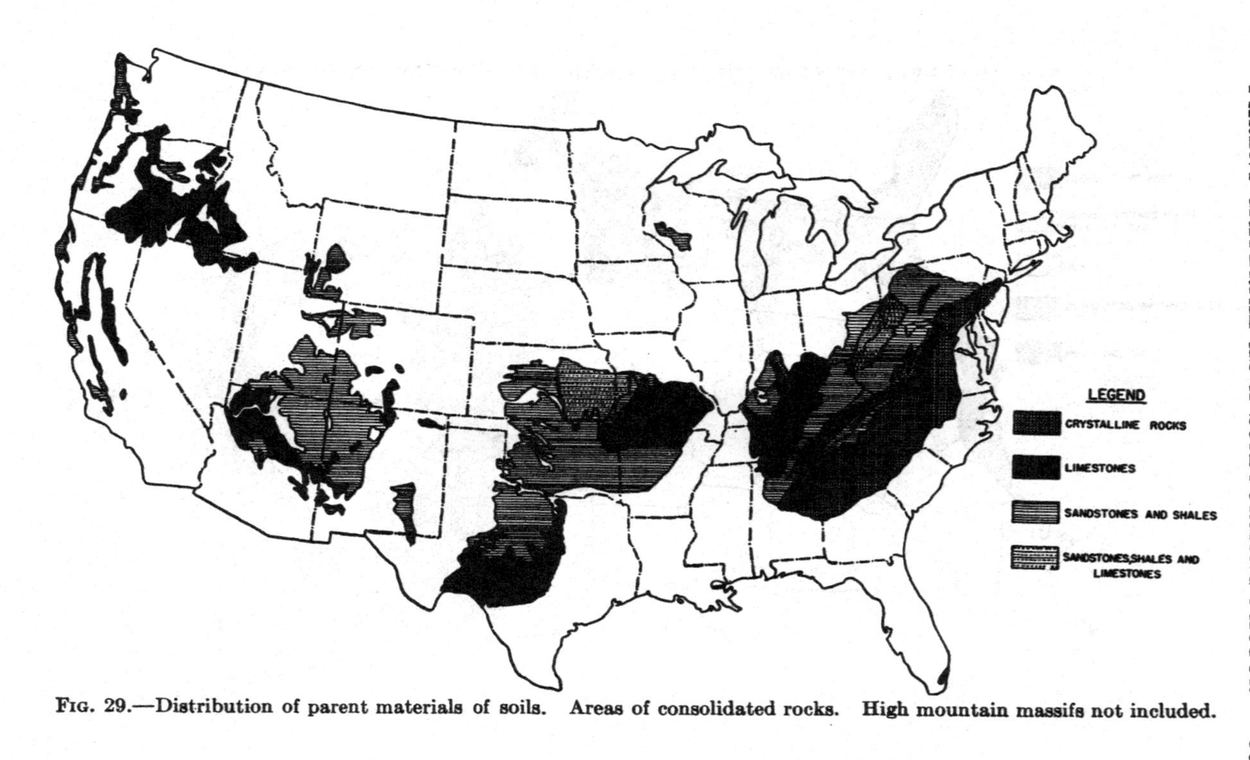

FIG. 29.—Distribution of parent materials of soils. Areas of consolidated rocks. High mountain massifs not included.

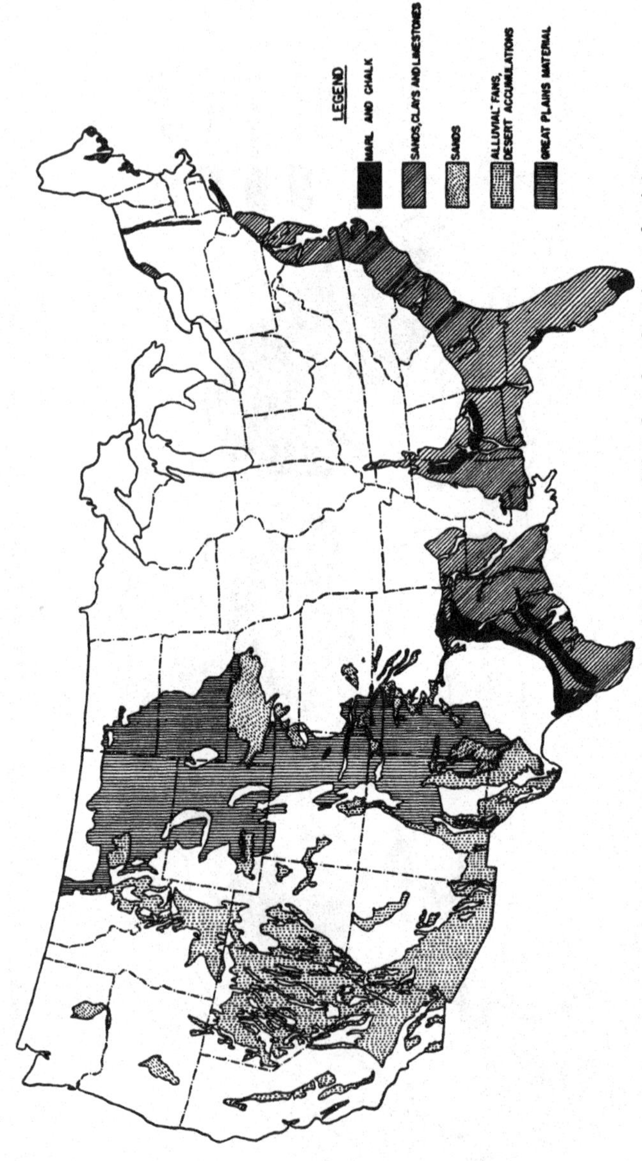

FIG. 30.—Distribution of parent materials. Unconsolidated rocks of variegated origin.

Physical and Chemical Rock Decay.—Experimental studies on the physical disintegration of rocks have been conducted by Hilger in Erlangen (compare page 32). He found limestone to be most resistant. Mica-schist particles having diameters of from 4 to 6 mm. rapidly broke into smaller fractions. Sandstones were the least stable materials.

Chemical rock decay has been repeatedly imitated in vitro. Daikuhara (5) leached powdered rocks with carbonated water for a period of 12 weeks. The order of decomposition was found to be as follows:

Basalt > Gneiss > Granite > Hornblende-Andesite

Optical examination is a common means of ascertaining relative degrees of weathering of minerals. Unfortunately, the reports found in the literature are not consistent, and it is difficult to formulate general rules. Fair agreement seems to exist for the following comparisons:

1. Quartz is one of the most resistant minerals.
2. Among the feldspars the plagioclases, especially their basic representatives, weather more readily than the potassium feldspars.
3. Biotites decay more rapidly than muscovites.
4. Amphiboles (*e.g.*, hornblende) are not so resistant as pyroxenes (*e.g.*, augites).

The fundamental difficulty of ascertaining relative rankings of chemical decay, even under controlled conditions, lies in the nature of the weathering process itself. It is a surface phenomenon, and the first step may be considered as an ionic exchange reaction

$$\boxed{\text{Mineral}}\begin{matrix}\text{K}\\ \text{K}\end{matrix} + \underset{\text{water}}{\text{HOH}} \rightleftarrows \boxed{\text{Mineral}}\begin{matrix}\text{H}\\ \text{K}\end{matrix} + \underset{\text{pH 9 to 11}}{\underset{\text{in solution;}}{\text{KOH}}} \qquad (11)$$

In the foregoing hypothetical potassium mineral, the K is exchanged for H. The resulting H surface is likely to become unstable (O^{--} of the lattice becomes OH^-, which tends to alter the bond strength), and the surface layer partly peels off. To arrive at quantitative orders of rock disintegration, it would be necessary to compare pieces with equal surface areas. Since we know of no reliable method to determine surface area of aluminosilicates, the weathering series mentioned above can have no claim to finality.

In nature, the extent of chemical rock decay may be very deceiving. Niggli's (20) decomposed gneiss of San Vittore-Lumino, Misox (T = 10.9°C., NS quotient = 624) has all the appearances of a completely weathered rock; yet chemical analysis can detect little change other than hydration and oxidation.

Significance and Behavior of Cations.—The nature of the elements released during rock decay has a specific bearing on soil formation. Silicon and aluminum furnish the skeleton for the production of clay colloids; iron and manganese are important for oxidation-reduction processes, and they strongly influence soil color; potassium and sodium are dispersing agents for clay and humus colloids, whereas calcium and magnesium have high flocculating powers and assure soil stability.

Acid igneous rocks contain considerable amounts of quartz and are rich in monovalent cations, whereas basic igneous rocks are high in calcium and magnesium contents. One would expect that these chemical and mineralogical differences would be reflected in the trend of soil formation. Indeed, Hart, Hendrick, and Newlands (9) in their studies on the soils of Scotland found that, under conditions of identical climate and topography, the basic igneous rocks produce brown earth and the acid igneous rocks produce podsolized soils.

It is a common saying that acid igneous rocks give rise to soils of good physical condition, whereas soils from basic rocks possess favorable chemical characteristics that ensure abundant plant growth. Aside from the fact that well-known exceptions exist (*e.g.*, serpentine soils), it is probable that such belief originated in regions where chemical weathering is not pronounced. In warmer climates with extensive leaching and removal of bases, the chemical influence of the parent rock is likely to be less marked. Cobb's work seems to support this conclusion.

Cobb (4) has published chemical analyses of igneous rocks and soils derived therefrom that were collected in the North Carolina section of the Piedmont Plateau. He arranged the data in the form of two development series as follows:

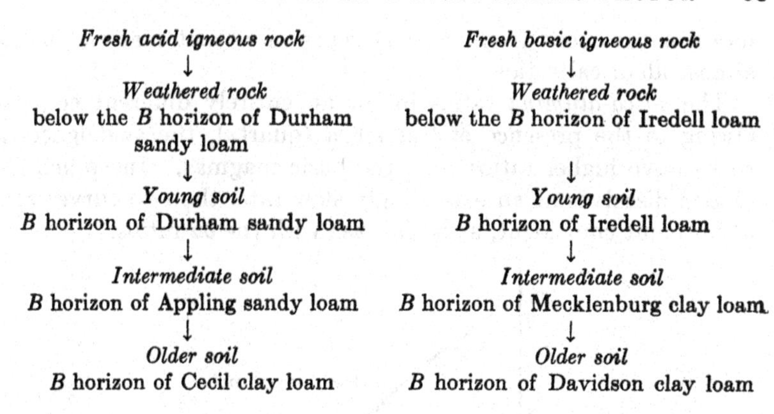

Cobb's analyses, recalculated to molecular values, are plotted in Figs. 31 and 32.

The *ba values*, or the base-alumina ratio that reflects the leaching of potassium, sodium, and calcium, and the relative accumu-

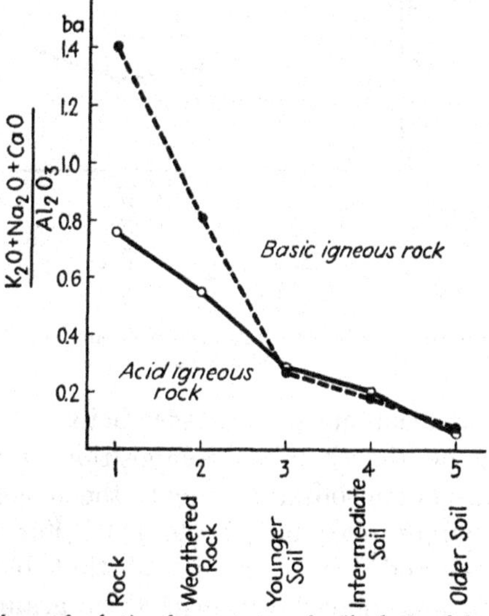

FIG. 31.—Comparison of relative base status of soils derived from acid and basic igneous rocks. (*Cobb's data.*)

lation of aluminum, show perceptible differences only in the earlier stages of weathering and soil formation. Basic igneous rocks and young soils derived from basic igneous rocks have higher *ba* values than the granites and gneisses. The older and more mature soils from various types of parent material have almost identical ratios.

The *silica-alumina* ratios follow an entirely different course. Owing to the presence of free silica (quartz), the acid igneous rocks have higher ratios than the basic magmas. Inasmuch as quartz dissolves at an exceedingly slow rate, the two curves are at no point coincident, as is the case with the *ba* ratios.

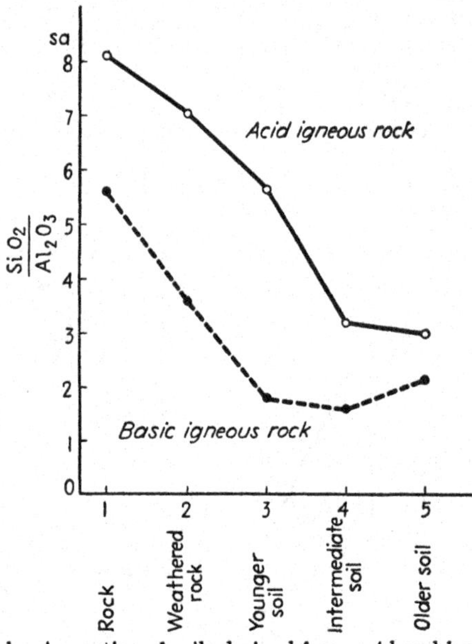

FIG. 32.—Silica-alumina ratios of soils derived from acid and basic igneous rocks. (*Cobb's data.*)

Chemical Composition of Surface Soils.—Jensen (12) has arranged data on the chemical composition (hydrochloric-acid extracts) of Australian soils according to the geological characteristics of the parent material (Table 11). For the two major groups, *granites* and *basalts,* he found that basaltic soils are higher in lime and phosphoric acid than granitic soils. The potassium and nitrogen contents could not be definitely correlated with parent material.

Composition of Rocks and of Soil Organic Matter.—Soil formation and development of vegetation are concomitant. As the soil features change, the plant cover adjusts itself accordingly. In very immature soils, the situation is different. Within a given climatic region, the growth of vegetation is mainly determined by the character of the parent material, whether limestone, igneous rock, sand deposit, or clayey shale. Not only type of vegetation but its chemical composition as well is affected. This influence is reflected in the decomposition products of the vegetational debris or, in other words, in the constitution of soil organic matter. A good example of this is given by Leiningen's (16) analyses of alpine humus soils (Table 12). Humus layers developed on igneous rocks are much lower in calcium, but higher in potassium and phosphorus than those of calcareous origin. The nitrogen content shows little difference between the two types of rocks.

TABLE 11.—EFFECT OF PARENT MATERIAL ON CHEMICAL COMPOSITION OF AUSTRALIAN SOILS (*Jensen*)
(Averages)

Region	CaO, per cent		P_2O_5, per cent		N, per cent		K_2O, per cent	
	Granitic soil	Basaltic soil	Granitic soil	Basaltic soil	Granitic soil	Basaltic soil	Granitic soil	Basaltic soil
Southern Tableland	0.125	0.306	0.100	0.226	0.149	0.125	0.122	0.273
West central Tableland	0.135	0.262	0.105	0.170	0.100	0.221	0.113	0.115
Northern Tableland	0.211	0.241	0.104	0.192	0.104	0.207	0.159	0.122

TABLE 12.—COMPOSITION OF ALPINE HUMUS DEVELOPED ON LIMESTONE AND ON ACID IGNEOUS ROCK (*Leiningen*)

Constituents	Limestone (nine samples), per cent	Igneous rock (five samples), per cent
CaO	2.75* (0.71–5.88)†	0.665 (0.17–1.19)
K_2O	0.053 (0.021–0.088)	0.171 (0.126–0.317)
P_2O_5	0.121 (0.045–0.187)	0.219 (0.155–0.264)
N	1.77 (1.46–2.29)	1.71 (1.37–2.00)

* Average.
† Extreme values in parentheses.

C. SOIL FORMATION ON SEDIMENTARY ROCKS

Alluvial deposits, moraines, loess, etc., are not necessarily soils. Unconsolidated sedimentary rocks are classified as soils only if at least one of their properties (*s* type) becomes a function of soil-forming factors.

Water Permeability of Parent Material in Relation to Soil Formation.—The downward movement of water is one of the prime factors in the transformation of a parent material into a soil with characteristic horizon differentiation. According to the regional concept of soil formation, all parent materials within a given environment should produce soils that ultimately possess similar morphological features. However, the speed with which the soil climax is reached varies enormously with the constitution of the soil matrix, particularly with its capacity for water percolation. If, for example, a sandy deposit and a silty sediment are placed in a podsol environment and inspected after a given interval of time, say 10,000 years, the former is likely to show a well-developed ortstein layer, whereas the latter may show only weak podsolization. The two profiles probably would be classified as belonging to different soil series.

Carrying the above reasoning a little further, it is conceivable that variations in water permeability not only affect the rate of development of the regional soil type, but also in extreme cases may give rise to the genesis of different climatic soil types and aclimatic profiles. The following arbitrary selection of parent materials may serve to illustrate this point.

Very Coarse Parent Material (Blocks).—Water runs through too rapidly to allow of profound chemical rock decomposition and translocation of weathering products. The soil climate is pseudoarid (Penck's term). Rock avalanches on mountain slopes fit into this group.

Fine Sandy and Silty Parent Material.—Water percolation is neither too fast nor too slow, resulting in a relatively rapid development of regional soil types.

Loamy Parent Material.—Movements of the liquid phase are slowed up, so that the transpiration stream of plants is able to pump out enough water to reduce materially the amount of percolating soil solution. This condition tends to shift the soil climate toward aridity. In northern Europe Scherf (26) observed

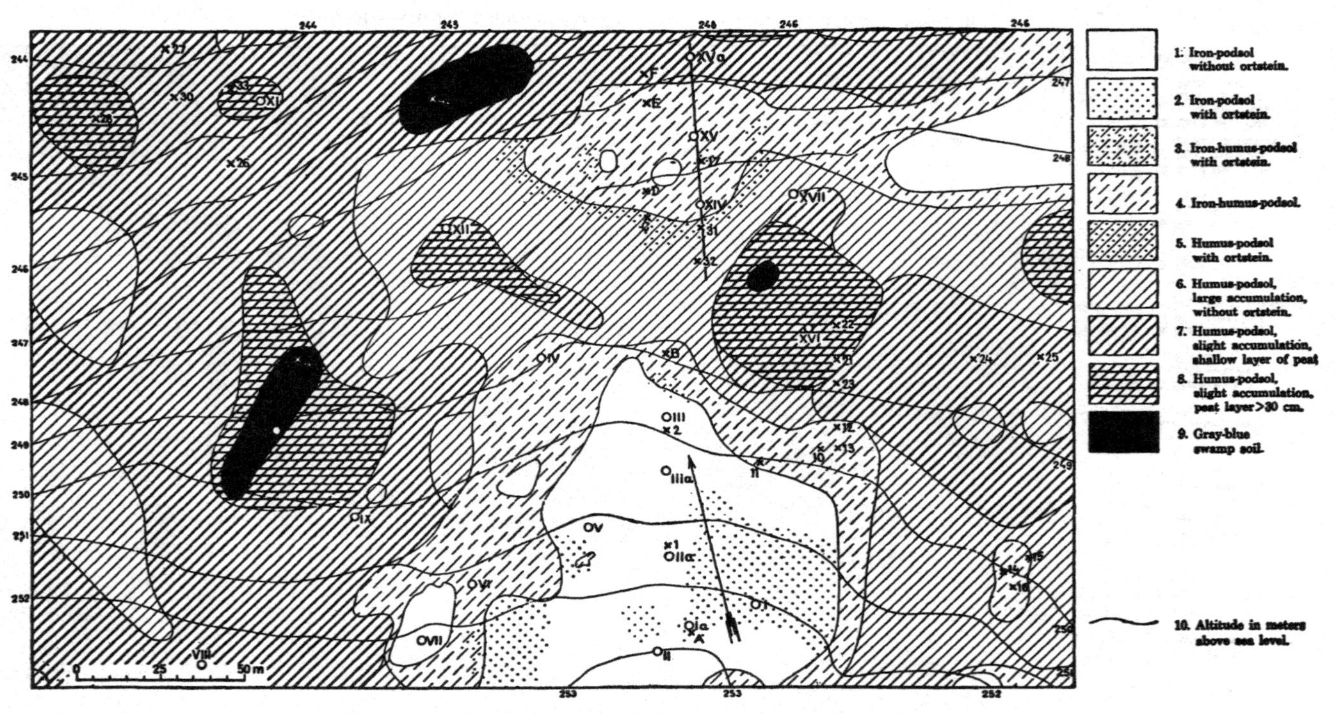

FIG. 33.—Soil map of the Rockliden Experiment Field in Sweden.

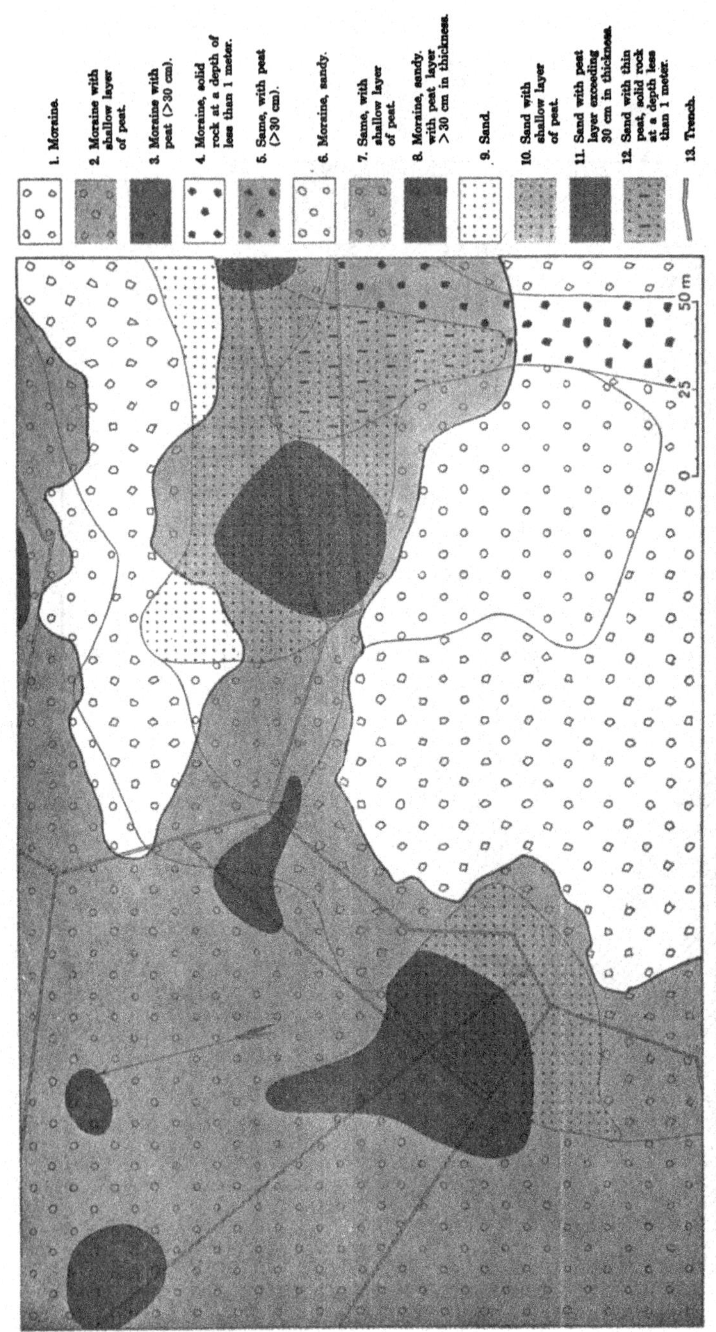

Fig. 34.—Geologic map of the Rockliden Experiment Field in Sweden.

brown forest soils on heavy ground moraines, whereas adjacent sands were strongly podsolized. Polynov (22) discusses the intrusions of podsols into the tundra zone and into the south Russian chernozem belt. In the latter region, wherever there exists a contact zone of loess and sand, steppe profiles are associated with the loess, whereas podsolized forest soils have developed on the sands. Similar physiographic complexes are found in central Oklahoma. Within extensive areas of chestnut soils and prairie soils, broad sand ridges emerge above the rolling plains. They are covered with forests and apparently are podsolized.

Clayey Parent Material.—Water penetration is strongly impeded, drainage is poor, and the soils are aclimatic. They lack characteristic regional profile features.

A classic example of the intimate relationship between parent material, drainage conditions, and soil formation is furnished by the Rockliden Experiment Field in northern Sweden (30). In accordance with the cool, humid climate of the locality, the regional profile is a podsol, but the area is by no means covered by a uniform soil type. The soil map (Fig. 33) shows a mosaic pattern of various types of iron podsols, iron-humus podsols, humus podsol, peat, and swamp soils. Tamm believes that the multitude of the profile differentiations is caused by the variable character of the vegetation, which, in turn, is conditioned by the properties of the parent material, particularly its water permeability and its relation to the ground-water table (Fig. 34). Where the substratum is sand or light-textured moraine, the ground-water level is low, at least from 60 to 70 cm. (23.6 to 27.5 in.) below the surface, and the vegetation consists of spruce forest with *Vaccinium*. The soils are iron podsols. On less permeable material, with higher ground-water levels, sphagnum plant associations cover the soil, and humus podsols prevail. Where the ground-water table reaches the forest floor and downward movement of water is nil, the soils are of a peaty nature or belong to the gray-blue swamp-soil group.

In conjunction with the researches of Tamm, it is of interest to list Hesselman's and Malmström's measurements of water percolation in glacial parent materials in their natural setting (Table 13). In heavy unweathered moraines, the flow of water is slow, only from 1.5 to 2.4 cm. (0.59 to 0.94 in.) per day.

Winters and Washer (32) have correlated qualitatively the water permeability of heavy Wisconsin moraines with the clay and colloid content of the parent material (Table 14). If the clay content ($<5\mu$) exceeds 60 per cent, the material is practically impervious to water.

TABLE 13.—RELATIVE PERCOLATION VELOCITIES OF WATER IN FLUVIO-GLACIAL SEDIMENTS (*Malmström*)

Material	Velocity (relative)
Unweathered moraine (heavy)	1
Moraine with humus podsol (A_2 horizon)	4–11
Moraine with iron podsol (A_2 horizon)	251
Sand	>3,500

TABLE 14.—WATER PERMEABILITY OF UNLEACHED GLACIAL DRIFT OF WISCONSIN AGE (*Winters and Washer*)

Permeability	Clay content $<5\mu$, per cent	Colloid content $<1\mu$, per cent	Soil type
Moderate to slow	37–50	18–27	Elliot
Slowly permeable	50–60	27–32	Plastic Elliot
Impermeable	>60	>32	Clarence

Peculiarities of Calcareous Parent Materials in Soil Formation.—In humid regions, the properties of soils formed from limestone rocks are related to the impurities in the parent material, especially the sand grains and the clay particles. Although these amount to less than 10 per cent for the average limestone, they accumulate rapidly during the process of soil development, since the carbonates are readily dissolved and carried away.

The relative accumulation of clay particles and siliceous pebbles in soils derived from calcareous parent materials is strikingly illustrated in the case of Midwestern moraines. Kay (13) has analyzed the petrographic constitution of Pre-Illinoian Pleistocene deposits in Iowa. The unleached portion of the drift is rich in pebbles, nearly half of them consisting of limestone and dolomite. During the interglacial periods, the surface of the moraines underwent pronounced weathering. Calcareous material dissolved, and the igneous rocks disintegrated and formed clay particles. The rapid disappearance of the limestone pebbles brought about a relative accumulation of insoluble silica rocks from 13 per cent to 78 per cent (Table 15). Correspondingly, the clay content became enriched to such an extent that the present surface of the moraine constitutes a clayey, tenacious, impermeable stratum known as "gumbotil."

TABLE 15.—PEBBLE ANALYSIS OF NEBRASKAN TILL (*Kay*)

Type of pebbles	Number of pebbles, per cent	
	In unleached drift	In gumbotil
Limestone and dolomite	44.6	0
Shale	1.0	0
Sandstone	1.9	0
Quartz	3.3	36.8
Chert, flint, etc	4.3	21.3
Quartzite	5.7	20.3
Schist	0.2	0
Granite	13.3	8.3
Basalt and greenstone	25.1	11.0
Unidentified or remaining	0.6	1.8
Total	100.0	99.5

A qualitative description of soil genesis on limestones, with special emphasis on the vegetational factor, will be offered in Chap. VII. On the quantitative side, the model investigation of Van Baren (31) deserves special consideration. The transformation of rock into soil can be easily traced from the data listed in Table 16. In the surface soil, the carbonates have nearly vanished, humus has been synthesized, and the finer particles ($<20\mu$) have accumulated from mere traces to over 20 per cent.

It is well known that in humid regions the presence of alkaline earth carbonates retards the regional soil-forming processes. Colloidal sols of $Al(OH)_3$ and $Fe(OH)_3$, when unprotected by colloidal humus, are readily flocculated by $CaCO_3$ solutions. The migration of clay particles within the soil profile is reduced. In consequence, the formation of illuvial and eluvial horizons is impeded, and only after the removal of the carbonates are the characteristic regional profile differentiations instigated. Limestones tend to produce immature soils, often designated as rendzinas or humus-carbonate soils.

TABLE 16.—FORMATION OF A LIMESTONE SOIL IN HOLLAND (*Van Baren*)
(Analyses by L. Möser, Giessen)
Annual temperature = 11.0°C. (51.8°F.)
Annual rainfall = 76.9 cm. (30.3 in.)
NS quotient = 313

Horizons	Horizon C_2, limestone rock	Horizon C_1, disintegrated rock	Horizon *B*, soil	Horizon *A*, cultivated surface soil
Depth	Below 230 cm.	170–230 cm.	30–170 cm.	0–30 cm.
SiO_2, per cent	5.16	25.58	65.73	73.38
TiO_2, per cent	0.06	0.23	0.69	0.93
Al_2O_3, per cent	0.15	3.56	8.57	7.59
Fe_2O_3, per cent	0.87	2.57	6.37	2.63
FeO, per cent	0.14	0.13	0.28	0.82
MnO, per cent	0.02	0.03	0.05	0.08
MgO, per cent	0.58	1.10	1.12	0.38
CaO, per cent	51.74	35.37	3.39	1.01
Na_2O, per cent	0.06	0.12	0.42	0.55
K_2O, per cent	0.20	0.33	1.43	1.38
P_2O_5, per cent	0.09	0.11	0.41	0.10
SO_3, per cent	0.04	0.04	0.06	0.05
CO_2, per cent	39.85	27.11	1.72	0.74
Cl, per cent	0.023	0.025	0.058	0.050
N_2O_5, per cent	0.001	0.001	0.001	0.001
H_2O above 110°, per cent	0.47	0.62	4.11	2.96
H_2O below 110°, per cent	0.42	2.88	5.07	3.09
Humus, per cent	—	0.13	0.40	4.42
Total per cent	99.874	99.936	99.879	100.161
pH	7.3	7.7	7.8	7.3
Particles less than 0.02 mm.	—	15.8	19.6	21.5
Identified minerals	18	15	11	13

Under identical climatic conditions, calcareous soils appear darker and are usually richer in organic carbon than the non-limestone soils. The calcium ions coagulate the humus colloids and thus prevent their dispersion and subsequent removal. Exceptions, however, are known. In the steppe regions of the U.S.S.R., soils derived from chalk are said to be lighter in color than those of loessial origin (Polynov). In Poland, Miklaszewski (18) has identified even white varieties of rendzina.

In certain parts of the world, especially in warm climates, the limestones produce yellow and reddish soils. Ramann (24) maintained, many years ago, that the lateritic soils of the tropics advance on limestone deposits into midlatitudes. The famous terra rossa of the Mediterranean coasts, and the reddish soils of the Ozark Mountains are often cited as examples.

It has been reported that the red color of limestone soils may depend upon the quality of the parent material. Hard and relatively pure limestone rocks frequently produce red soils, whereas the softer and impure varieties yield dark-gray and brownish weathering products. As examples of the first kind, Nevros and Zvorykin (19) cite the red patinas on the marble columns of the Temple of Jupiter in Athens, on the Arch of Hadrian, and on public buildings. Examples of the second kind are found on monuments in Athens and particularly in Delphi.

Laterites on limestones and dolomites supposedly have been observed in India and South Africa, but Harrassowitz emphatically denies the existence of laterites on limestones. He is of the opinion that limestone rocks produce aclimatic soils not only in the temperate regions but also in the tropics.

Effect of Lime on the Rate of Leaching of Potassium and Sodium. Ions held on surfaces of minerals and clay particles cannot be removed by leaching unless they are exchanged for other ions. In the presence of pure water, the release of potassium and sodium is accomplished by the hydrogen ions of the water in accordance with the following hydrolysis equation:

$$\boxed{\text{Particle}}\begin{matrix}K\\Na\end{matrix} + \underset{\text{Water}}{\begin{matrix}HOH\\HOH\end{matrix}} \rightleftarrows \boxed{\text{Particle}}\begin{matrix}H\\H\end{matrix} + \underset{\text{(In solution)}}{\begin{matrix}KOH\\NaOH\end{matrix}} \qquad (12)$$

Potassium and sodium released in this manner may be readsorbed during the subsequent migration within the soil mass.

Laboratory experiments (11) have demonstrated that $CaCO_3$ has a pronounced effect on the exchange of monovalent ions as indicated by the following general equilibrium reaction:

$$\boxed{\text{Particle}}\begin{matrix}K\\Na\end{matrix} + CaCO_3 \rightleftarrows \boxed{\text{Particle}}Ca + (K, Na)CO_3 \qquad (13)$$

Ca enhances the release of K and Na and, in addition, reduces their readsorption, because of effective competition. These effects are especially pronounced for sodium.

Jenny (10) has examined existing soil analyses for the purpose of determining whether or not the above Ca exchange principle also operates in nature. Analyses of podsolized soils from humid temperate and cold regions were grouped into two classes according to the presence or absence of calcium carbonate in the parent material.

Class I. Parent materials containing carbonates (>1 per cent).
- *a.* Shales, sands, moraines, loess,
- *b.* Limestones.

Class II. Parent materials without carbonates.
- *a.* Igneous and metamorphic rocks,
- *b.* Shales, alluvial sands.

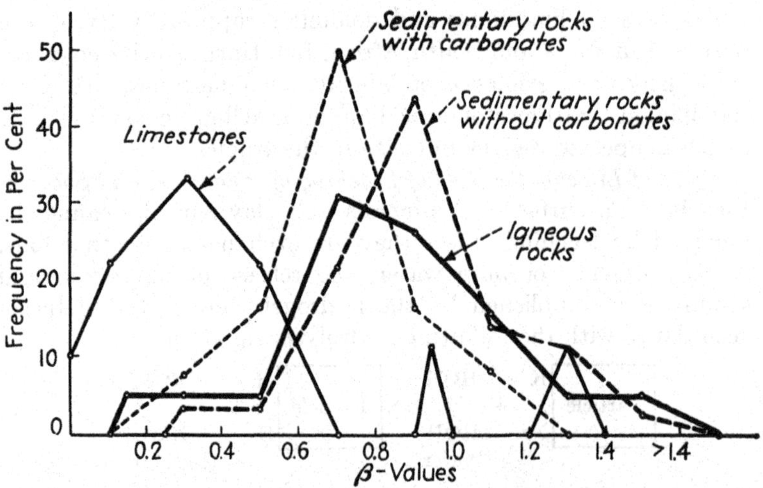

FIG. 35.—Frequency distribution of leaching values (β) of podsolized soils as affected by parent material.

For each of the 68 soil profiles the leaching factor β was calculated according to methods outlined on page 27. A graphical summary is presented in the form of a frequency diagram in Fig. 35. The leaching values are plotted on the abscissa, whereas the relative number of profiles that fall into arbitrary class intervals of 0.2β is shown on the ordinate. The effect of parent material is clearly manifested, especially in the case of limestones versus igneous and metamorphic rocks. Here, the modes, namely, the β values that correspond to the highest peaks in the curves, are 0.30 and 0.80, respectively. Enhanced leaching due to calcium carbonate is strongly suggested. There are a few low β values within the igneous rock group. They refer to profiles developed on basaltic rocks that are high in calcium, which, as one would expect, behave more like calcareous rocks. On the other hand, several β values exceed the magnitude 1, and it is possible that in these profiles alumina has been leached more than the monovalent bases, a conclusion that is not without foundation in regions of podsolic weathering.

Contrasting the two major classes of parent materials, the statistical averages assume the following magnitudes:

Parent materials with $CaCO_3$ (21 profiles), $\beta = 0.620 \pm 0.071$.

Parent materials without $CaCO_3$ (47 profiles), $\beta = 0.888 \pm 0.047$.

Again the calcareous soils have significantly lower β values. One is led to believe that the exchange principle outlined in the beginning of the discussion is indeed operative in weathering and soil formation and reflects itself in the analyses, in spite of many adverse effects, particularly biological influences.

The Loess Problem.—In many parts of the United States, particularly in the Mississippi, Missouri, and Ohio valleys occur large areas of a peculiar sediment known as "loess." Typical loess is a yellow-colored, unconsolidated deposit of silty texture, rich in calcium carbonate. It feels soft and tends to form upright bluffs; it is full of calcified capillary channels and contains shells of specific land snails, but no shells of aqueous species. Occasionally bones of land vertebrae (*Elephas*, *Mastodon*, *Equus*) are found. Sediments of similar nature, but with one or more of the above criteria missing, are called "loesslike" deposits. Loess without calcium carbonate is often called "loess loam."

Although the *origin of loess* is a geological rather than a pedological problem, the question of its formation has occupied the minds of all students of loessial soils. Some of the most carefully studied soil-formation series deal with loess.

Aquatic Theory.—The extensive loess deposits along the Rhine River in Europe and the Mississippi River in America led the earlier geologists to believe that loess must be water laid. Their theory was that during the Pleistocene many streams carried large amounts of mud that settled in quiet and shallow waters at times of large floods. However, the general absence of stratification and the occurrence of well-preserved land snails and bones of land vertebrae were somewhat difficult to explain by the aquatic theory. For these reasons, it has been largely abandoned, especially in western Europe and North America. In Russia, the alluvial theory still appears to have considerable stronghold, and there seems to exist good evidence of water-laid loess in the Volga and Dnieper regions. In recent years, Russell (25) has revived the aquatic theory of loess formation.

Aeolian Theory.—In 1877, von Richthofen advanced the theory that the immense loess sheets in China had been carried by the prevailing winds from the desert and dry steppe to the semihumid zones where the dust had become precipitated by rain and held to the ground by the vegetational grass cover. This viewpoint reconciles the absence of stratification and the presence of fossils of land fauna. In accordance with the aeolian concept, many scientists place the source of the Mississippi Valley loess in the western Great Plains area and the deserts of the Great Basin.

With the aid of detailed soil maps and many hundreds of deep borings, Smith and Norton (29) were able to construct a map that not only depicts the *distribution* of loess in the state of Illinois but also the *thickness* of the loess mantle (Fig. 36). They made the significant discovery that the depth of the loess deposits is directly related to the present major stream channels. The heaviest loess deposits are adjacent to the wide bottoms of the Mississippi River, as in the Mason County region, and in St. Clair and Clinton counties, bordering the Mississippi River. Smith and Norton refute the idea that the Illinoian loess has to any extent been blown from the Western plains and deserts. They attribute the source to the mud-laden rivers descending from the melting continental ice sheets. In periods of low-water levels during the Pleistocene, the wide, barren river bottoms were swept by winds that picked up the sediments and redeposited them on the uplands. Similar views were held by Fenneman (6).

Berg-Ganssen Hypothesis.—Berg (1) in Russia, and Ganssen (7) in Germany emphatically point out that a good loess theory should explain the universal uniformity of texture and the predominance of the silt-particle size. According to Berg, loess and loesslike deposits have a common origin: they are formed *in situ* through weathering of calcareous deposits in a dry climate. Ganssen advances the idea that, in regions of moderate rainfall,

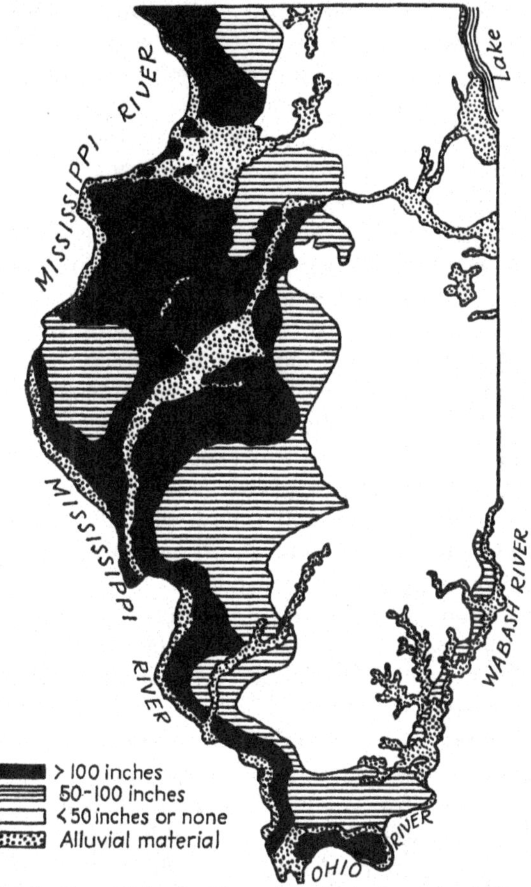

Fig. 36.—Distribution and depth of loess deposits in the state of Illinois. (*Smith and Norton.*)

physical rock disintegrations practically stop at the silt fraction, whereas the clay colloids of originally clayey deposits are aggregated to stable particles of silt size.

The Berg-Ganssen hypothesis of loess formation has received considerable publicity, but it is doubtful that the theory applies to the extensive loess deposits in the Middle West. In Missouri, for example, there are many striking examples of aeolian or even aquatic deposition of fine-textured strata on limestone rocks rich in chert and on moraines containing quartzite boulders.

Loessial Soils.—In conformity with the evenness of loess deposits, the loessial soils are exceedingly homogeneous. Within a given locality, the variability of the soil characteristics is small, especially if contrasted with morainic soils.

Since their deposition in the Pleistocene, the older loesses have undergone extensive weathering and soil formation. In southern Illinois and northern Missouri, clay contents as high as 40 to 60 per cent are frequently encountered. It is, however, somewhat problematic whether all this clay has been formed *in situ*.

Loessial soils are predominantly fertile, partly because of favorable chemical properties, partly because of excellent structural features. They acquired and preserve these properties by virtue of a moderate climatic environment in which they escape the destructive influences of extreme podsolization and laterization. Typical zonal soil types associated with loess are chestnut soils, chernozems, prairie soils, and to a lesser extent gray-brown podsolic soils, and yellow-red soils.

Soil Variability as Related to Parent Material.—Soils that appear uniform upon casual inspection in the field may reveal wide fluctuations in properties if subjected to chemical analysis. It seems that parent material is one of the important agents responsible for spatial variability of soil characteristics within a given soil type.

An impressive example is offered by the soil transect in Fig. 37, which demonstrates the variability of calcium carbonate of a cultivated brown forest soil of morainic origin. The samples were gathered at 30-ft. intervals. Within surprisingly short distances the lime content of the fine earth (<2 mm.) varies enormously, from 0 to 19 per cent calcium carbonate, as indicated by the upper, solid curve.

It can be easily shown that this variation in lime content is a direct reflection of the constitution of the moraine. The lower, dotted curve in Fig. 37 represents the number of limestone pebbles expressed as per cent of all stones. The parallelism of the two curves is striking, and there is little doubt that the calcium carbonate content of the fine earth is directly related to the petrographic make-up of the parent material of the surface soil.

Alkali soil transects are extremely irregular. From an experimental plot near Fresno, Calif., Kelley (14) collected samples 40 ft. apart. The graphs of total salt, sodium chloride, and sodium carbonate contents produce the complex pattern shown in Fig. 38. This sporadic variability also is reflected in the growth of vegetation. It is generally believed that the spotted nature of alkali soils is a consequence of the geologic origin of the soil matrix.

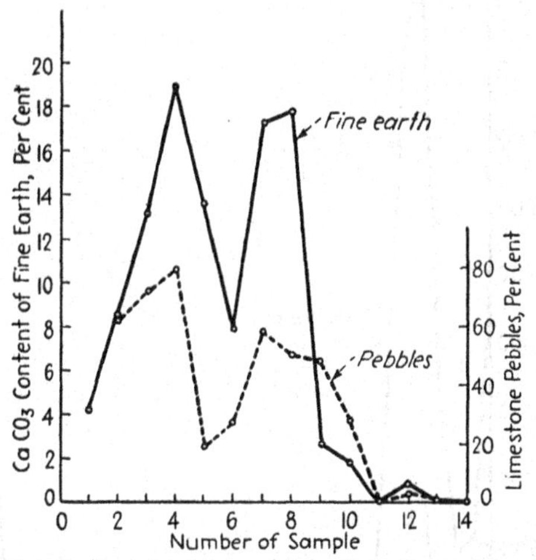

FIG. 37.—Areal distribution of calcium carbonate in the surface soil of a moraine. Upper curve: composition of fine earth; lower curve: composition of pebbles.

Pseudoprofiles.—A *genetic* profile is the result of the combined action of physical, chemical, and biological processes upon a given parent material. In particular, the horizons *A* and *B* owe their existence to vertical translocation of inorganic and organic substances within a geological stratum.

There are quite a number of profiles that are composed of well-defined layers that cannot be considered genetic horizons, although they are often erroneously labeled with the conventional letters *A*, *B*, *C*. To illustrate the point in question, attention is directed to Fig. 39, which portrays the profile of certain phases of the Lindley loam and Shelby silt loam. Under virgin timber, the *A* horizon is composed of a light grayish, silty material of uniform texture. The clayey *B* horizon is dark brown to reddish in color. The *C* horizon is composed of a mottled gray and yellow Kansan moraine. At first sight, the profile looks like a moderately developed podsol. Upon closer examination, one discovers that the C and B horizons are intermixed with pea- and nut-sized fragments of chert, quartzite, and amphiboles, whereas the A horizon lacks these essential features of morainic materials. It is completely free of stones. If we conclude that Lindley loam constitutes a truly genetic profile, we must assume that all stones

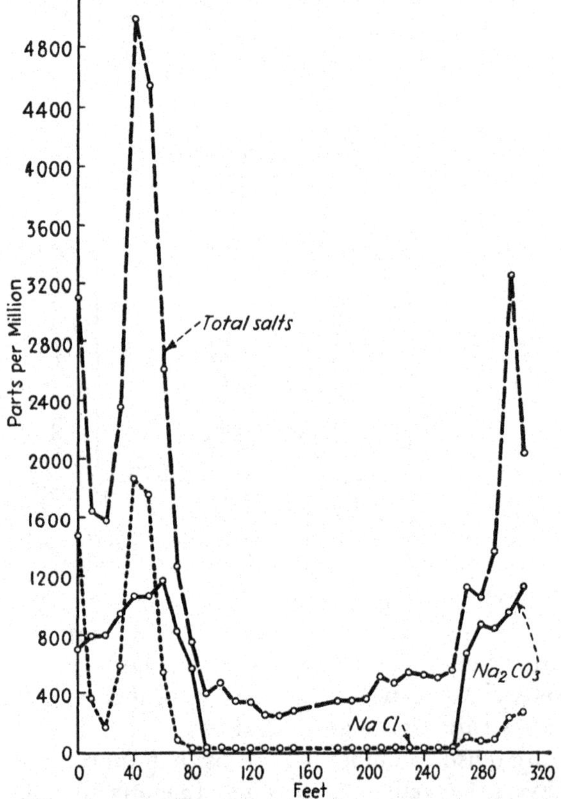

Fig. 38.—Graphs showing the linear variation of an alkali soil, first foot.

(flint, quartzite!) must have been dissolved in the A horizon, a postulate that is in conflict with the known observations of decay of minerals. The issue is further complicated by the fact that not infrequently the surface silt layer is disproportionately thick, varying from 1 to 3 ft.

A more rational explanation of the Lindley profile under discussion rests on the belief that it is not a truly genetic profile.

The *B* horizon most likely represents an old weathered glacial till that was exposed at one time. The *A* horizon may be considered to be a later geological deposit, probably a cover of loess. Thus in the pedological sense we are in all probability dealing with a pseudoprofile that owes its "horizons" to different geological strata.

The author is inclined to believe that the occurrence of pseudoprofiles is more widespread than is generally admitted. Recent investigations by Nikiforoff (21) on the solonetz-like soils of California seem to support this opinion.

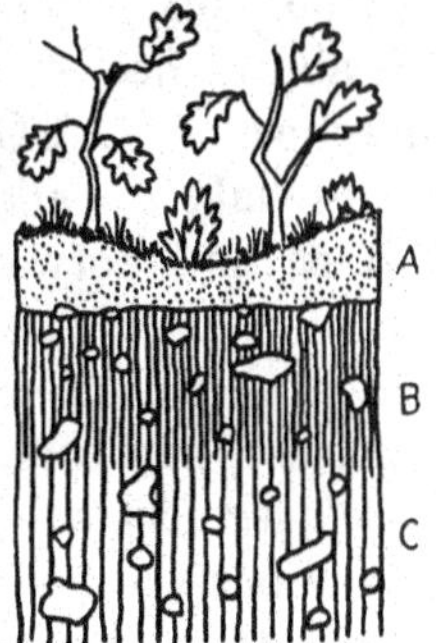

FIG. 39.—Schematic representation of a pseudoprofile.

Lee's Comparison of English and New Jerseyan Soils.—Lee (15) discovered that in England and in New Jersey the various soils types occur in banded areas that correspond in length, width, and general direction to those of the different geological formations. The texture of the soils is solely dependent on the nature of the parent rock. In central New Jersey, out of 48 soil types, 24, or 50 per cent, are light textured; in southeastern England, out of 57 types, only 14, or 25 per cent, are of sandy nature; the rest are loams and types heavier than loam.

The explanation of this lies in the prevalence of sandy unconsolidated geological formations in New Jersey, whereas in southeastern England heavy silty and clayey unconsolidated deposits and bedrocks containing, when composed, relatively high percentages of silts and low percentages of sands predominate.

In New Jersey all the surface and lower horizons are acid in reaction, whereas in southeastern England many soils are alkaline or neutral in the upper or lower horizons or both. This is a most striking relationship and points very clearly to the close relation between the reaction of the geologic materials and the soils. All the geologic materials in New Jersey are acid in reaction, and so are the soils, whereas in southeast England geological formations basic in reaction give rise to neutral or alkaline soils, and formations of acid reaction develop soils having an acid reaction.

Scherf's Geological Theory of the Formation of Alkali Soils in Hungary.—On the plains of Hungary, alkali soils, also known as "Szik" soils, are very extensive. Owing to their agricultural importance, they have been studied in great detail, especially by de Sigmond and his school. There are two types of alkali soils that deserve special attention.

Solontchak Soils.—These soils contain throughout the profile an abundance of water-soluble alkali salts. Because of the presence of $CaCO_3$, Na_2CO_3, and $NaHCO_3$, the Hungarian solontchaks are alkaline. The pH varies between 9 and 10 in the surface and between 8.4 and 8.6 in the lower horizons. The school of de Sigmond believes that the ground water, which is rich in alkali, at one time rose to the surface, alkalized the entire soil mass, and then receded.

Solonetz Soils.—In Hungary, solonetz soil has a salt-free, clayey surface horizon varying in thickness from about 8 in. to over 3 ft. The pH varies between 6 and 7.5. The physical condition of this horizon is poor, presumably as a consequence of the presence of sodium clay. Below the surface horizon, sharply set off, begins a heavy, alkaline subsoil (pH 8.5 to 10) with salt layers containing Na_2SO_4, $MgSO_4$, and $NaHCO_3$.

According to de Sigmond, the origin of solonetz may be explained as follows: first, the alkaline ground water rose to the surface, caused alkalinization, and then receded either naturally or artificially, just as in the case of solontchak. Subsequently, leaching took place, the surface soil became free of alkali, and the latter accumulated in the subsoil. A profile-development process was instigated, and the *A* horizons and *B* horizons of the solonetz profile were formed. On the basis of this picture, the solontchak soils represent immature solonetz soils.

Scherf (27) takes issue with the above soil-formation theory and maintains that the origin of solonetz has nothing to do with leaching processes; rather it is a consequence of peculiarities of the parent material. Scherf's findings and arguments are interesting and warrant closer examination.

In the first place, Scherf made a detailed soil-reaction map of a relatively small area. Over 3,000 pH measurements were taken, to a depth of 6 ft. With the aid of the pH data, he discovered a close relationship between soil reaction and nature of the geological substrata. Scherf then made 51 carefully leveled borings to depths of 21 ft. or greater, the distance between the holes varying from 6 to 450 ft. In this manner he was able to gain a detailed picture of the successions and the extent of the geological strata. For the region as a whole, the following significant layers were found, listed from the surface downward.

Holocene (10c). A fluviatile mud layer, free of calcium carbonate, originally acid (pH 4 to 6); no sodium; poor water permeability.

Uppermost Pleistocene (9a). A typical loess, aeolian deposit.

Uppermost Pleistocene (9b). Loesslike material, clayey silt of aeolian and fluviatile nature.

Uppermost Pleistocene (4 to 7). Fluviatile deposits, formed in swiftly running water. Mostly sand and some clay.

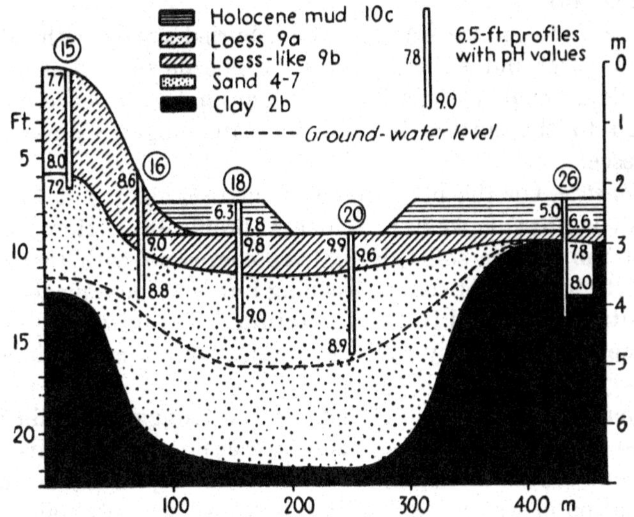

FIG. 40.—This graph illustrates the geologic theory of alkali soil formation in Hungary, according to Scherf.

Upper Pleistocene (2b). A dark, blue, green, or brown clay; contains calcium carbonate and is impermeable to water. A fluviatile deposit from slow currents.

The lowest stratum from the Upper Pleistocene (2b) was once a surface that was strongly eroded during the third interglacial period (Riss-Würm). The erosion channels and the elevations, although covered now with from 12 to 20 ft. of more recent material, can still be recognized and are clearly traceable by the deep borings (see black areas in Fig. 40). According to Scherf, these submerged valleys furnish the source of alkali water. The ground water, which itself contains some NaCl and Na_2SO_4, cannot escape the depressions, except through capillary lift into the surface, a process that is favored by the aridity of the Hungarian climate. As soon as the water reaches the present surface, it evaporates, and the salts accumulate in the topsoil. Thus, salty ground water near the surface and aridity are the necessary and sufficient prerequisites for the formation of alkali soils in Hungary.

Scherf further contends, and this is entirely novel, that the presence of solontchak or of solonetz is not related to leaching but depends on the nature of the geological stratum now exposed (see Fig. 40).

Solontchak.—If we stand above a buried valley with alkaline ground water and if the top layer "Holocene 10*c*" is absent, we have Solontchak (profile 20). The ground-water level is close enough to the surface so that the salts migrate upward and effloresce.

Solonetz.—On the other hand, if the originally acid Holocene deposit is present, we have solonetz (profile 18). Here, the low pH of the surface soil (pH 6.3), as compared with the subsoil (pH 9 to 9.8), is not due to leaching but is a consequence of the deposition of the acid Holocene stratum. The latter has, in part, become neutralized by the rising salts, and some sodium clay was formed.

Scherf claims that the climate of the Hungarian Plain is too arid to cause much leaching, and, furthermore, the texture of the Holocene deposit is too heavy to permit any significant downward percolation of water. According to Scherf, solonetz soils are not leached solontchak soils, as de Sigmond believes. They are not true profiles, because the horizons represent different geological deposits, namely, sand and loess in the *B* and *C* horizons, and fluviatile mud in the *A* horizon.

Figure 40 also shows the position of the regional soil types, the chernozem (profile 15). It occurs, in loess, above the submerged ridges of the buried upper Pleistocene (2*b*), where the capillary rise of the alkaline ground water is insignificant.

Scherf's theory has been quoted *in extenso*, not because we think it should definitely replace de Sigmond's ideas in general, but because it provides a striking illustration of the significance of careful differentiation between true soil horizons and geological strata. Whether or not Scherf's theory is the correct one will have to be decided in Hungary.

D. SYSTEMS OF SOIL CLASSIFICATION BASED ON NATURE OF PARENT MATERIAL

In the early days of soil science, the geologic or petrographic nature of parent material was taken as the sole basis for soil classification. With the advent of the climatic concept in classification, the geological systems were discredited, especially in countries that span entire continents, such as the United States and Russia. However, even today, in regions of relatively uniform climates and vegetation, the geologic classification criteria continue to enjoy wide popularity. The terms "glacial soils," "limestone soils," "loessal soils," and "basaltic soils," are widely used. Students interested in aspects of soil fertility cannot fail to notice the many ties existing between crop yields and geological strata. Among the gray-brown-podsolic soils, for example, are found the highly productive Hagerstown series in the limestone valleys of Maryland and Virginia and the submarginal Boone series developed on sandstones in the Mississippi River drainage system.

Polynov's System.—Polynov, although accepting the prevailing idea that the major soil groups, such as podsols and chernozems, are climatic types, maintains that they are composed of varieties that are essentially petrographic units. In regard to classification criteria, he points out that the rock-classification schemes devised by petrologists should not be blindly adopted. Variations within sedimentary rocks may produce soils that differ much more from each other than from igneous and metamorphous rocks.

Polynov's (22) classification of parent materials consists of the following major groups and subgroups.

I. Igneous rocks and crystalline schists.
   A. Acid rocks (gneiss, phyllite, mica schist, etc.),
   B. Neutral rocks,
   C. Basic rocks (pyroxenes, amphiboles, serpentine, etc.).

II. Transition forms.
   A_1., B_1., C_1. Deposits of the coarse disintegration products of Class I. Mainly gravel moraines, arkoses, gravel, and sandy alluvial deposits,
   A_2. Quartzites,
   B_2. Shales,
   C_2. Calcareous shales.

III. Sedimentary rocks.
   A. Quartz sandy rocks (sandstones, sands),
   B. Clayey rocks (fire clay, clay, loam),
   C. Carbonate rocks (limestones, dolomite, marl, etc.).

Shaw's System.—In the dry regions of California, intensive leaching rarely occurs, and because of this condition the petrographic and chemical composition of the parent material persists in the fully mature soil series.

In Shaw's (28) classification, the "order" is the most inclusive grouping; it is based on the nature of the soil mass, whether *mineral* or *organic.* The mineral soils are divided into *primary* (residual) and *secondary* (transported) soils. The "class" is the second subhead and is based on the *trend of reaction* in the soil mass (pedalferic and pedocalic soils). The "division," or third subhead, rests on those profile characteristics that are dependent on the petrographic nature of the soil material. Shaw established the five divisions:

Siallithous division. Soils derived from acid igneous materials (high in quartz).

Simalithous division. Derived from materials low in quartz (basic igneous rocks).

Arenalithous division. Soils derived from the weathering of sandstones and shales, with or without lime.

Calcilithous division. Soil derived from limestone rock sources.

Heterolithous division. Soils formed from sediments derived from mixed rock sources, mainly those deposited by major streams, with some extensive areas of wind-modified and wind-deposited soils that originally were water-deposited mixed sediments.

Generally speaking, soil classifications based on parent material assume special significance whenever the change from parent material to soil has not been profound. Such conditions prevail in regions of low rainfall (deserts) and low temperatures (frigid zones). Also, recent alluvial deposits and young soils in general belong to this group. If, on the other hand, weathering has been profound and leaching severe, the similarities between soil and parent material may be remote. Soils derived from limestones may be entirely void of calcium carbonate, and one may encounter the paradoxical situation that a farmer has to lime a "limestone soil." Under such circumstances, classification according to parent material is inadequate to portray the characteristic properties of a soil.

Literature Cited

1. BERG, L. S.: Loess as a product of weathering and soil formation, *Pedology (U.S.S.R.)*, **2**:21–37, 1927.
2. BROWN, P. E.: Soils of Iowa, *Iowa Agr. Expt. Sta., Special Rept.* 3, 1936.
3. BYERS, H. G., KELLOGG, C. E., ANDERSON, M. S., and THORP, J.: Formation of Soil, Soils and Men, *Yearbook of Agriculture*, U. S. Government Printing Office, Washington, D. C., 948–978, 1938.
4. COBB, W. B.: A comparison of the development of soils from acidic and basic rocks, *Proc. First Intern. Congr. Soil Sci.*, IV, 456–465, 1928.
5. DAIKUHARA, G.: Über saure Mineralböden, *Bull. Imperial Central Agr. Expt. Sta. Japan*, 2:1–40, 1914.
6. FENNEMAN, N. M.: "Physiography of Eastern United States," McGraw-Hill Book Company, Inc., New York, 1938.
7. GANSSEN, R.: Die Entstehung und Herkunft des Loess, *Mitt. Lab. Preuss. Geol. Landesanstalt*, Heft 4, 21–37, 1927.
8. HARRASSOWITZ, H.: Laterit, *Fortschr. Geolog. und Paleont.*, **4**: 253–566, 1926.
9. HART, R.: Soil mineralogy applied to problems of classification, *Trans. Third Intern. Congr. Soil Sci.*, **3**:161–162, 1935.
10. JENNY, H.: Behavior of potassium and sodium during the process of soil formation, *Missouri Agr. Expt. Sta. Research Bull.* 162, 1931.
11. JENNY, H., and SHADE, E. R.: The potassium-lime problem in soils, *J. Am. Soc. Agron.*, **26**:162–170, 1934.
12. JENSEN, H. I.: The soils of New South Wales, *Government Printer*, Sydney, 1914.
13. KAY, G. F., and APPEL, E. T.: The pre-Illinoian Pleistocene geology of Iowa, *Iowa Geol. Survey*, 34, 1929.
14. KELLEY, W. P.: Variability of alkali soil, *Soil Sci.*, **14**:177–189, 1922.
15. LEE, L. L.: Possibilities of an international system of the classification of soils, *J. South-East Agr. Coll.*, Wye, **28**: 65–114, 1931.
16. LEININGEN, W. ZU.: Über Humusablagerungen im Gebiete der Zentralalpen, *Naturw. Z. Forst- Landw.*, **6**:160–173, 1909; **10**:465–486, 1912.
17. MARBUT, C. F.: Soils of the United States, *Atlas of American Agriculture*, Part III, Washington, D. C., 1935.
18. MIKLASZEWSKI, S.: Mémoire relatif à la Pologne, *Mémoires sur la nomenclature et la classification des sols*, 245–255, Helsinki, 1924.
19. NEVROS, K., and ZVORYKIN, I.: Zur Kenntnis der Böden der Insel Kreta (Griechenland), *Soil Research*, **6**:242–307, 1939.
20. NIGGLI, P.: Die chemische Gesteinsverwitterung in der Schweiz, *Schweiz. mineralog. petrog. Mitt.*, **5**: 322–347, 1926.
21. NIKIFOROFF, C. C.: The Solonetz-like soils of southern California, *J. Am. Soc. Agron.*, **29**:781–796, 1937.
22. POLYNOV, B.: Das Muttergestein als Faktor der Bodenbildung und als Kriterium für die Bodenklassifikation, *Soil Research*, **2**:165–180, 1930.

23. Prescott, J. A., and Hosking, J. S.: Some red basaltic soils from Eastern Australia, *Trans. Roy. Soc. South Australia*, **60**: 35–45, 1936.
24. Ramann, E.: "Bodenkunde," Verlag Julius Springer, Berlin, 1911.
25. Russell, R. J.: Physiography of Iberville and Ascension Parishes, Louisiana Department of Conservation, *Geol. Bull.* **13**:1–86, 1938.
26. Scherf, E.: Über die Rivalität der boden-und luftklimatischen Faktoren bei der Bodenbildung, *Annal. Inst. Regii Hungarici Geol.*, **29**: 1–87, 1930.
27. Scherf, E.: Geologische und morphologische Verhältnisse des Pleistozäns und Holozäns der grossen ungarischen Tiefebene und ihre Beziehungen zur Bodenbildung, insbesondere der Alkalibodenentstehung, *Relationes Annuae Inst. Regii Hungarici Geol. Pro* 1925–28:1–37, Budapest, 1935.
28. Shaw, C. F.: Some California soils and their relationships, Syllabus JD, University of California Press, 1937.
29. Smith, R. S., and Norton, E. A.: Parent material of Illinois soils. In Parent materials, subsoil permeability and surface character of Illinois soils, 1–4, Illinois Agricultural Experiment Station and Extension Service, Urbana, 1935.
30. Tamm, O.: Studien über Bodentypen und ihre Beziehungen zu den hydrologischen Verhältnissen in nordschwedischen Waldterrains, *Medd. Statens Skogsförsoksanstalt*, **26**, 2, Stockholm, 1931.
31. Van Baren, J.: Vergleichende mikroskopische, physikalische, und chemische Untersuchungen von einem Kalkstein- und einem Löss-Bodenprofil aus den Niederlanden, *Mitt. Geol. Inst. Landbouw. Wageningen*, **16**:11–82, 1930.
32. Winters, E., and Washer, H.: Local variability in the physical composition of Wisconsin drift, *J. Am. Soc. Agron.*, **27**:617–622, 1935.
33. *Yearbook of Agriculture*, 1938 (Soils and Men), Washington, D. C.

CHAPTER V

TOPOGRAPHY AS A FACTOR IN SOIL FORMATION

Physiographers and geomorphologists have no generally accepted definitions of topography and relief. In the present discussion, the terms are used synonymously and denote the configuration of the land surface. Of the topographic designations commonly employed in pedology, the following are prominent: level or flat, undulating, rolling, hilly, and mountainous.

Topography as a soil-forming factor has not received the attention it deserves. It is true, of course, that a considerable amount of information on runoff and erosion in relation to slope is at hand, but it deals primarily with the removal and the destruction of soil and not with soil formation. It is mainly this latter aspect that this book undertakes to discuss.

Effect of Relief on Water Penetration of Soils.—A lucid exposition of the effect of relief on water penetration and profile formation is contained in Ellis' book (3) on the soils of Manitoba. With well-drained upland soils, level topography gives rise to soil-moisture conditions which, according to Ellis, are "normal" for the region. Soils under such conditions may be considered as representative of the regional climate. On the other hand, local variations in topographical position, such as knolls, slopes, and depressions, will result in moisture conditions that differ from the norm (Fig. 41). For example, if the precipitation on a given section of land is 18 in. annually, the soils on the knolls will receive 18 in., less the amount that runs off. Hence, the soils on the knolls will have a locally arid climate in comparison with the soils on level topography. Such soils may be designated as being "locally arid associates." The extent of this local aridity will be determined by the amount of water penetrating into the soil and the amount of runoff.

The soils of the depressions, in the examples given, will receive 18 in. of precipitation annually plus the amount of water that runs off from the adjacent higher lands. Hence, more water will penetrate the soils in the depressed areas, and such soils may be termed "locally humid associates," because they have a more humid soil climate than the soils on flat topography.

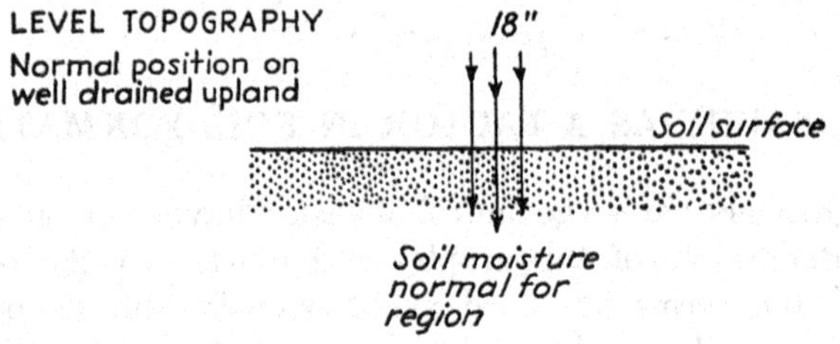

KNOB AND BASIN TOPOGRAPHY

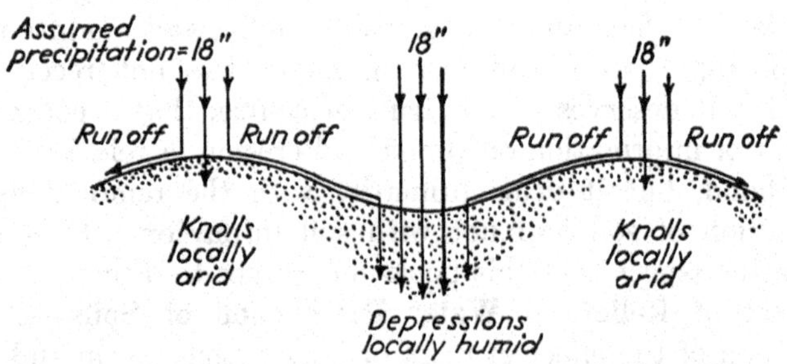

Fig. 41.—Effect of relief on water penetration of soils. (*Ellis.*)

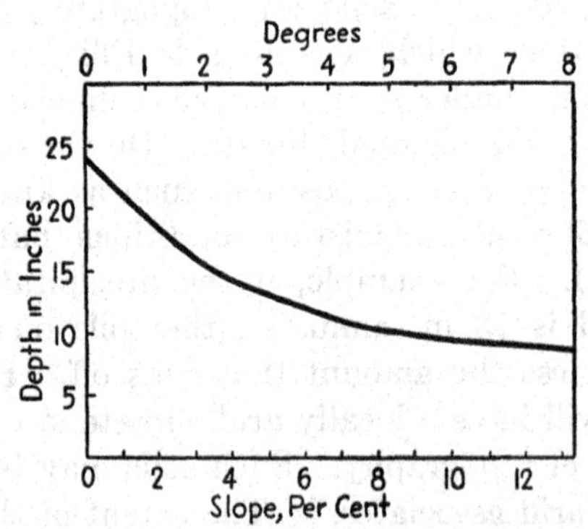

Fig. 42.—Relationship between slope and thickness of *A* horizon of timbered soils, derived from loess. The horizon has the greatest thickness on level topography and the least on steep slopes. (*Norton and Smith.*)

The profile differentiations resulting from these moisture differences are elucidated in the following sections.

Quantitative Relationships.—Functional analysis of the relief factor demands that the following equation be fulfilled:

$$s = f(r)_{cl,o,p,t,\ldots} \tag{14}$$

All factors, except topography, should be kept constant. Few systematic and quantitative investigations are on hand, although good opportunities exist for solving Eq. (14).

In general field practice, a soil type is accorded a certain kind of topography, such as undulating or hilly. In harmony with Eq. (14), a more refined approach necessarily would recognize a different soil type for each degree of slope. Thus, along a variable slope, there would be encountered an entire sequence of soil types, each having slightly different profile features. As an illustration, the work of Norton and Smith (12) on the forested loessial soils of Illinois may be quoted. These investigators made a great number of measurements of slopes and correlated the data with the depths of the *A* horizon. The average trend of the relationship is shown in Fig. 42. On flat areas, the thickness of the surface soil is 24 in.; on steep slopes, it is only 9 in.

TABLE 17.—RELATIONSHIP BETWEEN TOPOGRAPHY AND LEACHING IN LOESSIAL SOILS OF ILLINOIS (*Bray*)

Depth, inches	Constituents	Rolling topography (Clinton)	Level topography (Rushville)
at 10	Per cent clay, $<1\mu$	19.2	18.2
	pH	5.80	4.66
at 20	Per cent clay, $<1\mu$	30.0	38.4
	pH	5.44	4.70
at 30	Per cent clay, $<1\mu$	30.4	31.0
	pH	5.06	5.74
at 40	Per cent clay, $<1\mu$	28.7	26.6
	pH	5.00	6.90
at 50	Per cent clay, $<1\mu$	23.4	21.4
	pH	5.80	7.30
at 60	Per cent clay, $<1\mu$	19.4	19.8
	pH	6.60	7.76

In the same region, Bray (2) has obtained data on the relationship between slope and degree of leaching. His results are given in Table 17. The Clinton silt loam has a slightly rolling topography, whereas the Rushville silt loam developed under flat surface conditions. Greater leaching is found on the flat topography. The level soil (Rushville) is more acid in the surface portion and has more clay in the *B* horizon (from 15 to 30 in.).

The conclusion that profile formation is enhanced on level topography as compared with slopes should not be generalized without giving due consideration to the position of the ground-water table. We shall examine this restriction in the following sections.

Ground-water Table as a Soil-forming Factor.—In the introductory chapter, the fundamental equation of soil-forming factors was obtained by replacing the soil climate *cl'* with the air

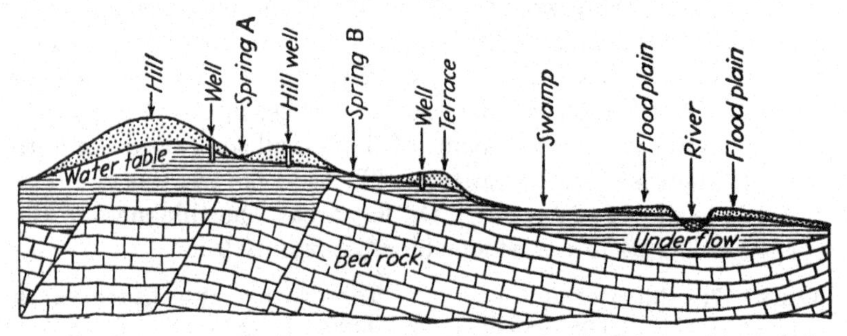

FIG. 43.—Ideal section across a river valley, showing the position of the ground-water table with reference to the surface of the ground and bedrock. [*After Slichter* (14).]

climate *cl*, on the assumption that *cl'* and *cl* are functionally related in such a way that the value of one is uniquely determined by the value of the other. This is true only when the ground-water table is so low that it does not influence the moisture of the soil. The extent of capillary rise from a free-water surface ($pF = -\infty$) depends on the texture of the soil. Generally speaking, where the ground-water table is more than from 8 to 12 ft. below the surface, its effect on the *A* and *B* horizons is slight and may be neglected.

In practical soil science, the influence of the ground-water table is recognized in the classification of level topography as flat uplands and flat lowlands (6). In theoretical discussions, it is deemed preferable to treat the ground-water table as an independent variable or soil-forming factor, because it can be made to vary independently of *cl*, *o*, *r*, *p*, and *t*. In humid climates, the following two cases merit special consideration.

Level Surface.—As may be seen from Fig. 43, the water table is the higher the greater the distance from a drainage channel.

The horizontal distance may be a matter of yards in cases of drainage ditches or of miles in the case of a major drainage channel such as a river.

Rolling Surface.—Under conditions of rolling topography, the depth to the water table increases as the distance from the draws becomes greater. Capillary rise may wet the surface of the depressions but not the crests of the ridges.

Specific examples of the relationships between ground-water table, topography, and profile features will be presented in the following sections.

Norfolk and Related Series.—Marbut has placed great emphasis on the effect of drainage on the characteristics of soil profiles.

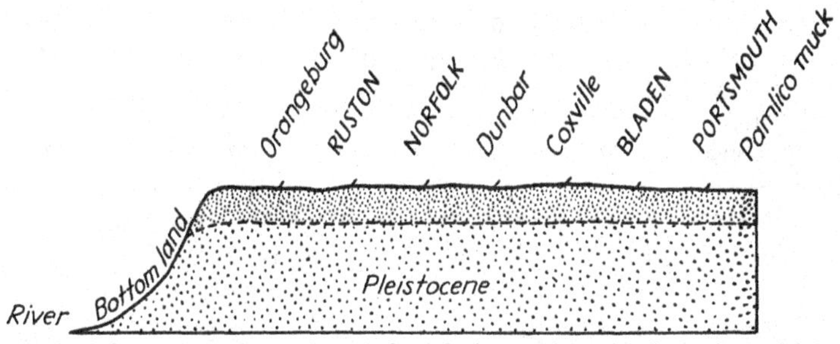

FIG. 44.—Schematic presentation of the relative position of the soils of the Norfolk group with reference to the distance from the main drainage channel.

The term "drainage," as used by pedologists, refers to the movement of water in the soil. It is conditioned by the precipitation, the configuration of the surface of the soil, the position of the ground-water table, and the permeability of the soil. The last, also spoken of as internal drainage, is a soil property and not a soil-forming factor.

Upon the initiative of Marbut, the former Bureau of Chemistry and Soils has studied the relationship between drainage and profile features for seven closely related soil series of the southern Atlantic Coastal Plain. The *C* horizons of the Ruston, Norfolk, Bladen, and Portsmouth series consist of unconsolidated sedimentary material and are of similar textural composition, especially in regard to the clay content. It may be assumed that the profile features of the above soil series are to a large, extent functions of the ground-water table, since the topography is uniformly flat and the climate is nearly identical throughout the region under consideration (Kinston, N. C.). The relative position of the various soils with respect to the larger drainage systems may be obtained from Fig. 44. The Ruston and

TABLE 18.—ANALYTICAL DATA OF THE NORFOLK AND RELATED SERIES

Soil types, fine sandy loams	Organic-matter content of *A* horizon (0–12 in.), per cent	pH		Base status $\frac{Na_2O + K_2O + CaO + MgO}{Al_2O_3}$	
		A horizon	Average of *B* + *C* horizons	*A* horizon	Average of *B* + *C* horizons
Good drainage:					
Ruston	0.88	6.0	4.9	0.185	0.040
Norfolk	0.89	5.1	4.6	0.121	0.026
Poor drainage:					
Bladen	2.74	4.4	4.3	0.060	0.038
Portsmouth	7.00	4.1	4.35	0.150	0.086

Norfolk series, which lie adjacent to the main drainage channels, have a low ground-water surface; whereas the Bladen and Portsmouth series, which are located in the more remote areas, have a prevailingly high-water table. A compilation of analytical data obtained by Holmes, Hearn, and Byers (5) is presented in Table 18.

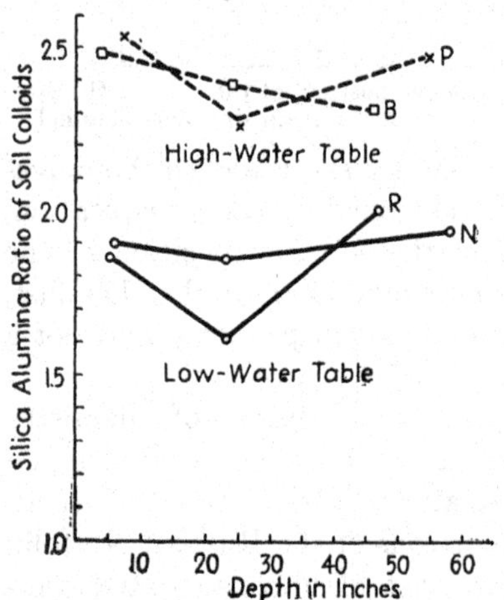

FIG. 45.—Effect of position of water table on the silica-alumina ratio of soil colloids (Norfolk group).

Organic matter and total nitrogen are high in the poorly drained soils, undoubtedly on account of anaerobic conditions developed during wet periods. The base status of all profiles is very low, and, accordingly, the acidity is high, as revealed by the low pH values. Soils formed in the presence of high ground-water table tend to be more acid than the profiles of the well-drained series. Whether this difference is due to variations in the base status or in the kinds of clay minerals cannot be elucidated from the data on hand. The latter possibility is suggested by the magnitudes of the silica-alumina ratio of the colloidal fraction (Fig. 45). Assuming that the composition of the clay particles at the beginning of the cycle of soil formation was identical, it follows that the better drained soils suffered a relatively greater loss of silica than the poorly drained soils. Likewise, the translocation of clay particles from the *A* horizons to the *B* horizons is very pronounced for the series with low ground-water tables. The ratio $\frac{\text{clay in } B \text{ horizon}}{\text{clay in } A \text{ horizon}}$ is 3.81 and 3.24 for the Ruston and Norfolk series, respectively, but only 1.21 and 2.27 for the Bladen and Portsmouth soils, respectively. In general, the study of soil formation on level topography indicates a diminution of profile differentiation under conditions of high ground-water tables.

Hydrologic Podsol Series.—The podsol profile is the result of leaching under a cover of sour humus. Where the water table is low, a light-gray A_2 horizon and a rusty-brown *B* horizon are developed (iron podsol). Where the water table is high enough to influence directly the pedogenic processes, the humus podsols are formed. In those extreme cases where a complete submergence of the mineral horizons occurs, nonpodsolized bluish-gray bog soils are formed. Mattson and Lönnemark (9) have studied a complete hydrologic podsol series in the rolling country near Lake Unden in Sweden. The entire sequence of soils occurs within a distance of 5 to 6 yd. The dry end of the series ends in a sandy hill that is several yards high and not affected by the water table. The wet end of the series terminates in a moderately wet depression covered with water-loving mosses. It is, however, completely submerged only during the rainy season. The vegetation of the entire sequence of soils consists of pine and spruce forest with a variable ground vegetation; *Polytrichum commune* with some *Sphagnum* is on the wet end, and *Calluna* and *Vaccinium* are on the dry end of the series. The parent material is a fine sandy glacial drift. Mattson's graphical presentation of the analytical data in relation to topographic features is quite novel and is reproduced in Figs. 46 and 47. The experimental values are plotted in a reference frame, the origin of which is at a point in the air above the wet end and in level with the dry end of the soil transect. The distances are measured in meters and decimeters.

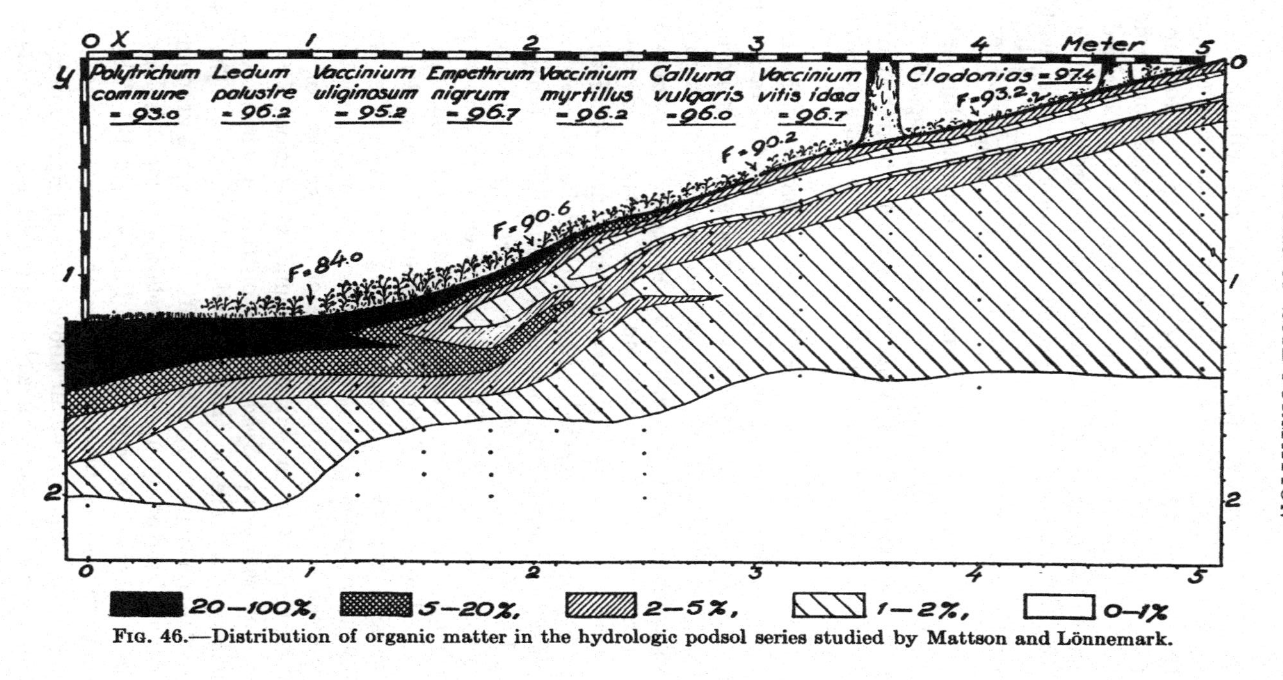

FIG. 46.—Distribution of organic matter in the hydrologic podsol series studied by Mattson and Lönnemark.

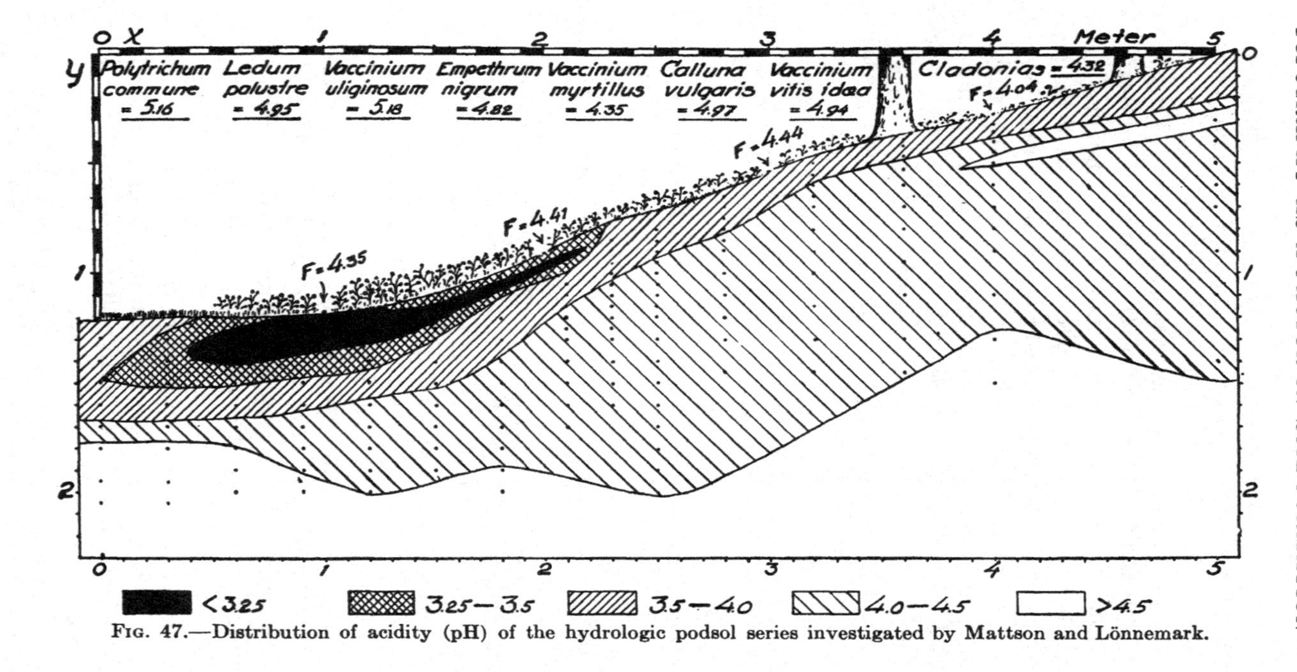

FIG. 47.—Distribution of acidity (pH) of the hydrologic podsol series investigated by Mattson and Lönnemark.

Owing to the small amount of clay present in the profiles, the *loss on ignition* may be taken as a measure of the organic-matter content of the soil. Starting from the origin ($x = 0$) the first 120 cm. are occupied by the peat-podsol zone. From $x = 120$ to $x = 210$ we find humus podsols that have a dark brownish-gray *B* horizon caused by a marked accumulation of humus. The iron podsols begin at $x = 230$ and extend to the end of the transect. Their *B* horizons are yellowish brown and are relatively low in humus. Equally illuminating is the *soil-reaction* chart. The pH determinations were made in suspensions containing 10 cc. of soil and 20 cc. of distilled water. The entire soil sequence is quite acid. The lowest pH recorded is 3.03 ($x = 90$, $y = 130$). It is of interest to note that the highest acidity occurs at the foot of the slope but not at the wettest portion of the series, which is at $x = 0$. High pH are found in the *C* horizons and in the *B* horizons of the iron podsols on the higher portions of the transect. Figure 47 also indicates the pH values of the aerial parts of the living plant materials. These are less acid than the dead plant parts that were collected on the surface of the profiles (*F* layers).

The hydrologic podsol series provides a good illustration of profile changes as a function of slope and drainage. Such sequences of changing profiles have been designated by Milne (10) as *catenas*.

Ground-water Table in Relation to Alkali Soils.—In arid regions, the ground water frequently contains amounts of dissolved salts sufficient that capillary rise may cause salinization of the soil profile. Naturally, soils on slopes and higher elevations tend to be less affected by this process than soils located in depressions. A good illustration of this is provided by the distribution of alkali in the soils of the Fresno family (13) of California, which are derived from uniform parent material of alluvial origin. The areas investigated comprise those of Visalia, Pixley, Wasco, Bakersfield, and Kings County in the southern part of the San Joaquin Valley. The data in Table 19 indicate what percentage of the area of the various soil series contains no alkali (free = less than 0.20 per cent), slight amounts of alkali (0.20 to 0.40 per cent), moderate amounts (0.40 to 1.00 per cent), and high amounts (more than 1.00 per cent) of alkali.

The Pond and Fresno series occupy the valley floor and are rich in alkali. Of the entire area of the Fresno soils, 93 per cent is highly infested with salt. None of the soils is free of alkali. The Cajon and the Traver series on the gently sloping alluvial cones have a sporadic distribution of salinity. Forty-one per

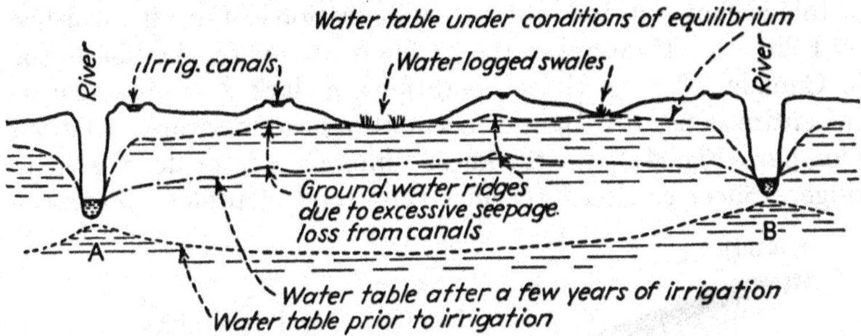

FIG. 48.—Illustrating the rise of ground-water table as a result of irrigation. Vertical scale greatly exaggerated. (*After Etcheverry* (4).)

cent of the area covered by Cajon soils is free of alkali, and only 22 per cent of the area contains excessive amounts of salt.

TABLE 19.—SALT CONTENT OF THE FRESNO FAMILY OF SOILS
(Data compiled by D. F. Foote from unpublished maps of the Soil Survey of the U. S. Department of Agriculture and the University of California)

Soil series	Topography	Distribution of salinity, per cent of total area			
		No alkali present	Slight alkalinity	Moderate alkalinity	High alkalinity
		(<0.20%)	(0.2–0.4%)	(0.4–1.0%)	(>1.0%)
Cajon....	Slope	41	22	16	22
Traver...	Slope	10	34	37	19
Pond....	Valley floor	0	3	29	67
Fresno...	Valley floor	0	0	7	93

The rise and fall of the ground-water table may take place quite independently of the age or maturity of the soil profiles. In certain sections of the plains of the San Joaquin Valley of California, the ground-water table has been raised artificially by irrigation (compare Fig. 48) as a result of which both immature and mature soils have suffered salinization. Some of the soils have become worthless for agricultural purposes. They produce no crops unless they are reclaimed.

Truncated Profiles.—Under a dense stand of grass or forest, sheet erosion is virtually nonexistent. However, when vegetation is removed and careless practices of cultivation are instigated, "man-made" erosion invariably results. Drastic alterations in profile characteristics may take place. An extreme example of this is shown in the schematic illustration of Fig. 49, furnished by Ellis (3). It describes the profile features of a chernozem soil in Canada. Under virgin conditions, a dark *A* horizon covers the entire slope, although, on account of local moisture variations (compare Fig. 41), its thickness diminishes from depression to ridge. Sheet erosion intensifies this trend of profile differentia-

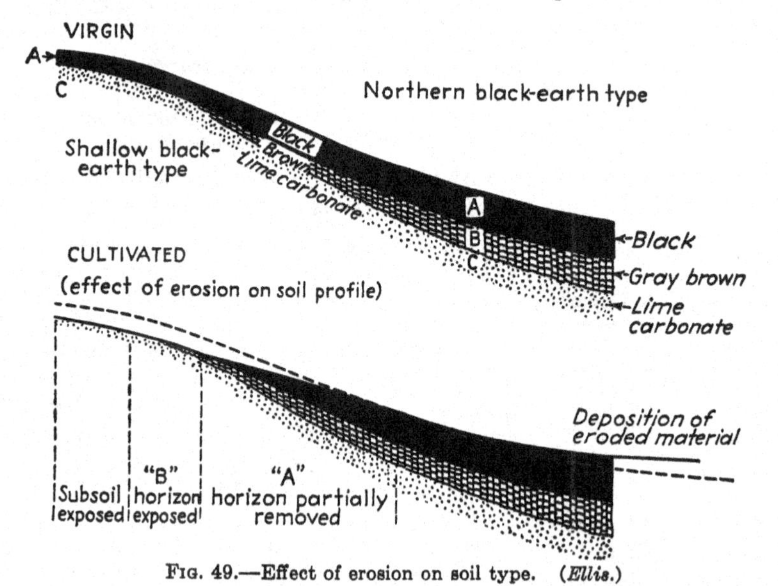

FIG. 49.—Effect of erosion on soil type. (*Ellis.*)

tion. Material is removed from the upper portions of the slope and deposited on the lower parts. Toward the ridges the *A* horizons may completely disappear, leaving the *B* horizon exposed. Such profiles are said to be truncated. Owing to differences in color of the *A* and *B* horizons, truncated profiles are often readily discernible on freshly plowed slopes. An example of widespread truncation is provided by the Cecil series of the Piedmont Plateau. The Cecil sandy loam possesses a gray, brownish-gray surface horizon that overlies a stiff red clay. Where the soil is not properly protected by a cover of vegetation, the torrential downpours during the early part of the spring carry away the light-colored sandy *A* horizon and expose the red-colored clayey *B* horizon. The soil is then mapped as Cecil clay. This truncated soil type comprises an aggregate of several million acres (1).

Marbut's Normal Relief and Normal Soil.—Marbut (7) writes:

> Experience has shown that in every region having what may be defined as *normal relief* there is a normal profile. By normal relief is meant the relief that at the present time characterizes the greater part of the earth's surface and may be described as smooth, undulating, or rolling, with a relation to drainage such that the permanent water table lies entirely below the bottom of the *solum*.

Flat topography and steep slopes produce soils without normal profile features, because "these profiles do not develop under an unimpeded and free action of the two dynamic soil-developing factors of the region, natural vegetation and climate." Marbut goes into considerable detail in his discussion of profiles in relation to relief, and, because of the importance of these concepts in relation to Marbut's classification of soils, a brief summary is presented herewith.

A.—Normal Profile.—It is formed exclusively on smooth, undulating topography. "The profile features are of the same *kind* as those in all the other soils of the same region and *all* the [profile] features are present which are present in any of the associated soils." Marbut cites the Marshall and the Clarion soils in the prairie region of Iowa as typical representatives of soils with normal profiles.

B.—Soils without Normal Profiles.—These are subdivided by Marbut into two types.

Type 1.—All (profile) features are present, but one or possibly more have developed to an excessive degree. These profiles occur on flat topography or on smooth slopes, but never on steep slopes. Examples: Clay-pan soils of the Middle West (Putnam, Cisni, Parsons) that are characterized by very heavy *B* horizons.

Type 2.—Some or nearly all the features of the normal soils are missing or imperfectly developed. This condition is related to slope in such a manner that the greater the slope—as compared with the "normal" slope—the fewer the number of characteristic profile features and the more feeble their design. Soils of mountainous regions furnish the most conspicuous examples of this type.

C.—Abnormal Profiles.—No feature is present which is identical with any feature of the normal soil. Their development has not followed the normal course. This group of soils is found predominantly in basins or in other poorly drained spots. Alkali and salty soils belong to this group.

Marbut has published a map (*Atlas of American Agriculture*, Plate 6) that shows the distribution and extent of the soils without normal profiles in the United States. Casual inspection of the map indicates that only about half the soils of the United States possess normal profiles.

It is probable that Marbut's concept of normal soils in relation to normal topography is a consequence of his lifelong interest in geomorphology. The idea of normal soils is helpful in visualizing the relationships existing between soil formation and cycles of geologic erosion, a point that also has been elaborated by Neustrujew (11). It should be made clear, however, that Marbut treats relief as a dependent variable, stressing its change as a function of time. Abnormal relief will eventually become normal relief. Soils never reach maturity unless they are associated with normal topography, according to Marbut. This view readily explains the profound differences in opinion prevalent among pedologists regarding the degree of maturity of certain soil types. In the prairie region, the Putnam silt loam is distinguished by a very heavy *B* horizon (clay pan), whereas the Marshall silt loam exhibits only feeble horizon differentiation. Local soil authorities regard the Putnam silt loam as the mature soil of the region into which the Marshall silt loam ultimately will develop. Marbut takes the opposite viewpoint. He regards the Marshall silt loam as the prototype of a mature prairie soil, because it has an undulating surface. The Putnam silt loam, on the other hand, occupies flat regions and therefore, according to Marbut, represents only a transitory state in the evolutionary process. In the course of time, the Putnam silt loam will become a Marshall silt loam. A similarly peculiar situation is found in California. The San Joaquin loam, which has an indurated hardpan, is not a normal soil as viewed from Marbut's standpoint; but, in the terminology of Shaw (13), it is representative of a mature profile.

Literature Cited

1. Bonsteel, J. A.: The Cecil clay, U. S. Department of Agriculture, Bureau of Soils, *Circ.* 28, Washington, D. C., 1911.

2. Bray, R. H.: Unpublished data.
3. Ellis, J. H.: The soils of Manitoba, *Manitoba Economic Survey Board*, Winnipeg, Manitoba, 1938.
4. Etcheverry, B. A.: "Land Drainage and Flood Protection," McGraw-Hill Book Company, Inc., New York, 1931.
5. Holmes, R. S., Hearn, W. E., and Byers, H. G.: The chemical composition of soils and colloids of the Norfolk and related soils series. U. S. Department of Agriculture, *Tech. Bull.* 594, Washington, D. C., 1938.
6. Kellogg, C. H.: Soil Survey Manual, U. S. Department of Agriculture, *Misc. Pub.* 274, Washington, D. C., 1938.
7. Marbut, C. F.: A scheme for soil classification, *Proc. First Intern. Congr. Soil Sci.*, Vol. 4, 1–31, 1928.
8. Marbut, C. F.: Soils of the United States, *Atlas of American Agriculture*, Part III, Washington, D. C., 1935.
9. Mattson, S., and Lönnemark, H.: The pedography of hydrologic podsol series, I, *Ann. Agr. Coll. Sweden*, **7**:185–227, 1939.
10. Milne, G.: A provisional soil map of East Africa, East African Agricultural Research Station, Amani, Tanganyika Territory, 1936.
11. Neustrujew, S. S.: Böden und Erosionszyklen, *Ref. Peterm. Mitt.*, **72**: 31, 1926.
12. Norton, E. A., and Smith, R. S.: The influence of topography on soil profile character, *J. Am. Soc. Agron.*, **22**: 251–262, 1930.
13. Shaw, C. F.: Some California soils and their relationships, Syllabus JD, University of California Press, 1937.
14. Slichter, C. S.: The motions of underground waters, U. S. Geological Survey, *Water-supply and Irrigation Papers*, 67, Washington, D. C., 1902.
15. Weir, W. W.: Shape of the water table in tile drained land, *Hilgardia*, **3**:143–152, 1928.

CHAPTER VI

CLIMATE AS A SOIL-FORMING FACTOR

Quantitative functional analysis is only possible if the soil property and the conditioning factors investigated can be expressed in numerical terms. The factor climate is so complex that no single numerical value can be assigned to a given climate. It becomes necessary to work with individual climatic components, the most important of which are moisture (m) and temperature (T). Treating these two subfactors as independent variables we may write the approximate equation

$$s = f(m, T)_{o,r,p,t,\ldots} \qquad (15)$$

This equation may be split into two formulas, one expressing *soil properties as functions of moisture*

$$s = f(m)_{T,o,r,p,t,\ldots} \qquad (16)$$

and the other expressing *soil properties as functions of temperature*

$$s = f(T)_{m,o,r,p,t,\ldots} \qquad (17)$$

The letter s denotes a soil property. Again, the subscripts indicate that the nature of the relationship between s and m or T varies with specific constellations of o, r, p, and t. For the sake of brevity the two relationships will henceforth be designated as soil property-moisture functions and soil property-temperature functions.

A. MOISTURE AS A SOIL-FORMING FACTOR

1. Discussion of Moisture Criteria

Rainfall.—Climates have been divided into arid and humid, the former referring to regions with scanty, the latter to regions with abundant rainfall. The differences in amount of precipitation over the earth are great. Yuma, Ariz., records a mean annual rainfall of 3.35 in., Mobile, Ala., one of 62.55 in., whereas at Mount Waialele, Kauai, Hawaiian Islands, an average of

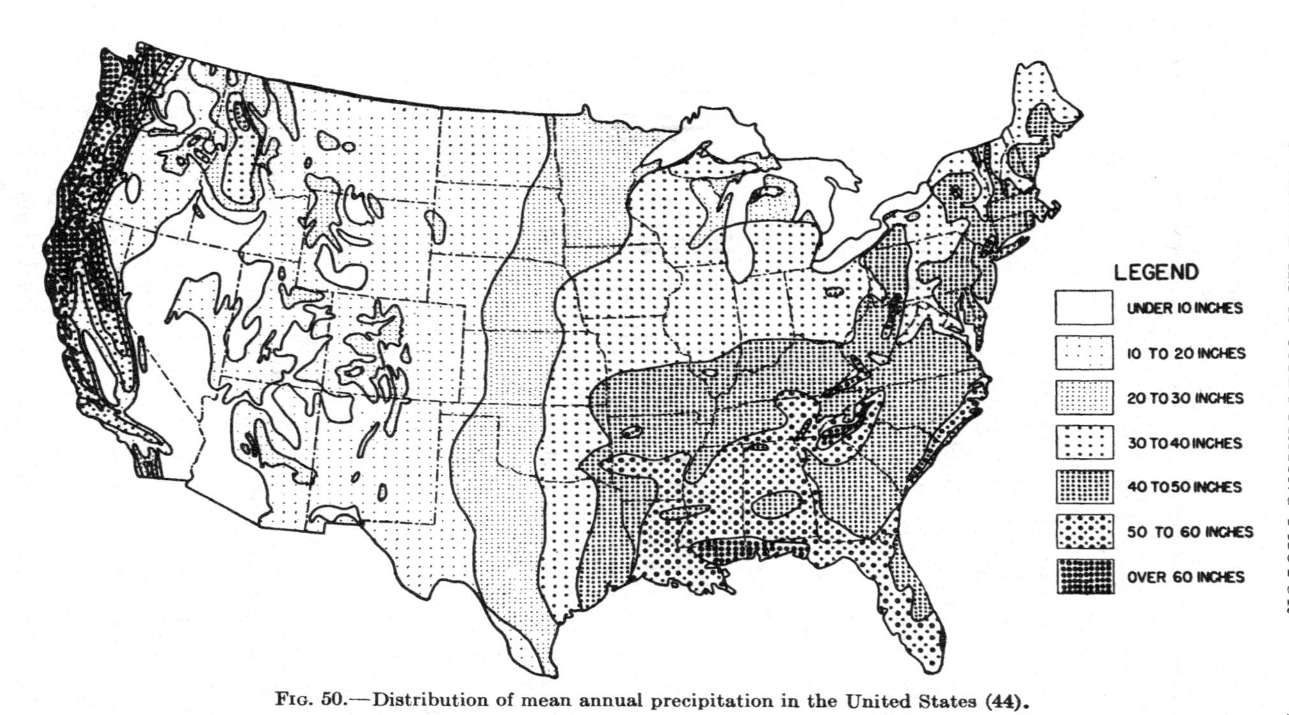

FIG. 50.—Distribution of mean annual precipitation in the United States (44).

476 in. falls annually. Such differences in rainfall profoundly affect the aspect of the landscape, especially the type of vegetation and the nature of the soils. A generalized rainfall map of the United States is shown in Fig. 50.

Annual values of precipitation provide a means for rapid characterization of the main moisture features of a region. Often *seasonal variations* are taken into account. Among the seasonal rainfall types that have been recognized in the United States, the Eastern, Plains, and Pacific are outstanding (Fig. 51). The Eastern type is characterized by a comparatively uniform distribution of precipitation throughout the year. The Plains

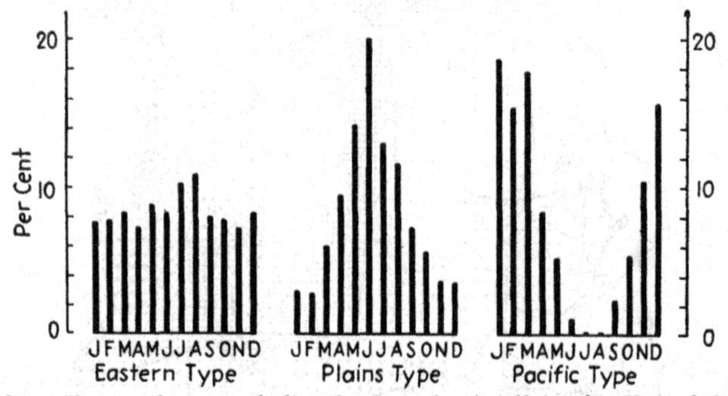

Fig. 51.—Types of seasonal distribution of rainfall in the United States.

type has dry winters and a marked concentration of rainfall in the late spring and summer months. In the region of the Pacific type, the winters are wet, and the summers are very dry. Theoretically, in any attempt to correlate soil features with annual moisture figures, attention to seasonal fluctuations must be given. The seasonal distribution pattern assumes the status of an additional independent variable or soil-forming factor. In practice, it has been found that minor variations in seasonal trends are of subordinate influence in soil formation.

As a result of soil-erosion studies, scientists are paying special attention to the *number and size of torrential rains.* The Agricultural Experiment Station at Columbia, Mo., (54) supplies the following information: during 14 years (1918–1931) of erosion experimentation, 420 rains caused runoff. Each of twenty-eight of these rains brought two or more inches of precipitation in 24 hr. These few but relatively heavy rains were responsible for over half the soil erosion that occurred during the entire period. In certain tropical regions, as much as 30 in. or more of rain may fall within 24 hr.

Evaporation and Transpiration.—Only a part of the precipitation that falls upon a level surface percolates through the soil profile. Much of the moisture evaporates or is given off by transpiration through plants. An examination of the evaporation map published by the U. S. Weather Bureau in the *Atlas of American Agriculture* shows that in the warm season alone, April to September, evaporation from a free-water surface reaches values from 25 to 88 in.

Unfortunately, evaporation from a free-water surface cannot be directly compared to evaporation from a barren soil or a vegetational cover. Various means to overcome this difficulty have been devised. The atmometer cup (Livingston) deserves special attention but cannot yet be considered a final step in the imitation of evaporation from soils. According to the extensive researches of Veihmeyer (84) and associates, the losses by transpiration may greatly exceed the losses by evaporation.

The Precipitation-evaporation Ratio.—Penck (56) has used precipitation and evaporation as a basis for his classification of climates. He has set the boundary between arid and humid regions at that locality where precipitation (P) and evaporation power (E) are equal and thus arrives at the following general groups:

E greater than P: arid regions,
E equal to P: arid-humid boundary,
E smaller than P: humid regions.

"E greater than P" means that the capacity of a region to evaporate water exceeds the actual precipitation.

As early as 1905, Transeau (80) constructed a precipitation-evaporation ratio map of the Eastern United States. The ratios were based on Russell's evaporation measurements from a free-water surface made during the year 1877–1878. The effective moisture zones are quite different from the rainfall zones, as is evident from a comparison of Figs. 50 and 52.

The main advantage of a precipitation-evaporation ratio map over a rainfall map lies in the possibility of comparing conditions of soil moisture of regions having different temperatures and different air humidities. For instance, St. Paul, Minn., and San Antonio, Tex., have about the same mean annual precipitation,

27.40 and 27.70 in., respectively. Yet the effective moisture conditions in regard to plant growth and soil formation are by no means alike. Actually, San Antonio's climate is much drier than that of Minnesota's capital. This difference is clearly indicated by the Transeau ratio, which is 0.51 for San Antonio and 1.02 for St. Paul. In accordance with these values the climatic soil-profile features of the two localities exhibit marked differences.

The Rain Factor.—As a substitute for the precipitation-evaporation ratio, which is difficult and laborious to determine experi-

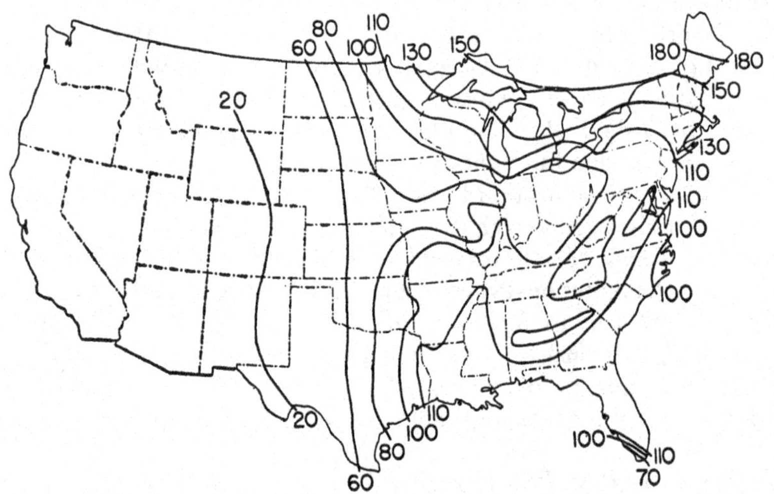

FIG. 52.—Distribution of Transeau's precipitation-evaporation ratio in the United States (80).

mentally, Lang (46) suggested the precipitation-temperature ratio (millimeters: degrees centigrade), or rain factor, in which evaporation is replaced by temperature. Its value in soil studies has been much disputed. A world map of rain factors has been published by Hirth (29), a more detailed map for the United States by Jenny (35). Attempts to correlate soil features with rain factors have been undertaken in Europe, Palestine (61), and the United States. In the main, they have not been very successful. Difficulties arise when the annual temperature falls below 0°C. because the ratio assumes a negative value. De Martonne (50) avoids this complication by dividing annual precipitation by "annual temperature + 10." Written in equations, the two values take the following form:

$$\text{Lang's rain factor} = \frac{P}{T}$$

$$\text{De Martonne's } \textit{indice d'aridité} = \frac{P}{T + 10}$$

P represents annual precipitation in millimeters, and T annual temperature in degrees centigrade. Hesselman (27) finds for Sweden a satisfactory agreement between the distribution of vegetation and de Martonne's *indice d'aridité*. Interestingly enough, Angström (2) discovered a close correlation between

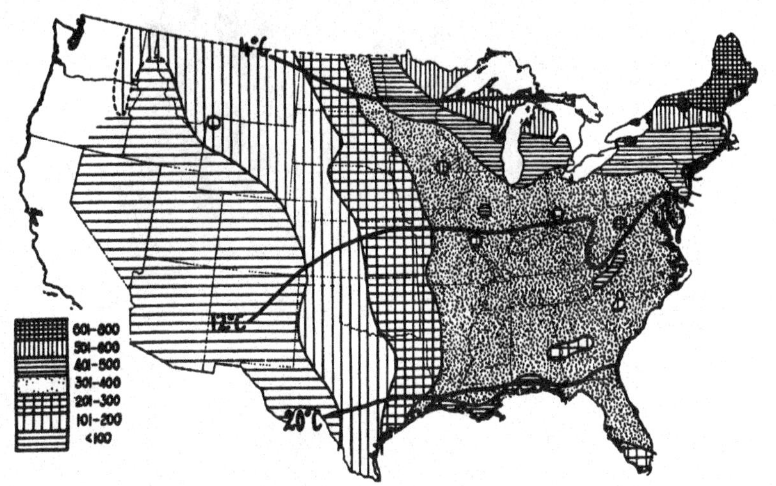

FIG. 53.—Moisture zonation of the United States according to annual *NS* quotients. Mountainous regions and Pacific Coast not included. The heavy black lines indicate mean annual isotherms.

monthly factors of de Martonne and duration of precipitation as expressed in minutes per month.

The NS Quotient.—A widely used substitute for the precipitation-evaporation ratio is Meyer's (52) *NS* quotient, which is obtained by forming the ratio

$$\frac{\text{Precipitation (millimeters)}}{\text{Absolute saturation deficit of air (millimeters mercury)}}$$

In the United States, it parallels the Transeau ratio quite closely over a wide moisture range. In Australia, Prescott (58, 59) observes a constant relationship between evaporation and saturation deficit that permits the calculation of the Transeau value by dividing the *NS* quotient by the factor 230. *NS* quotient

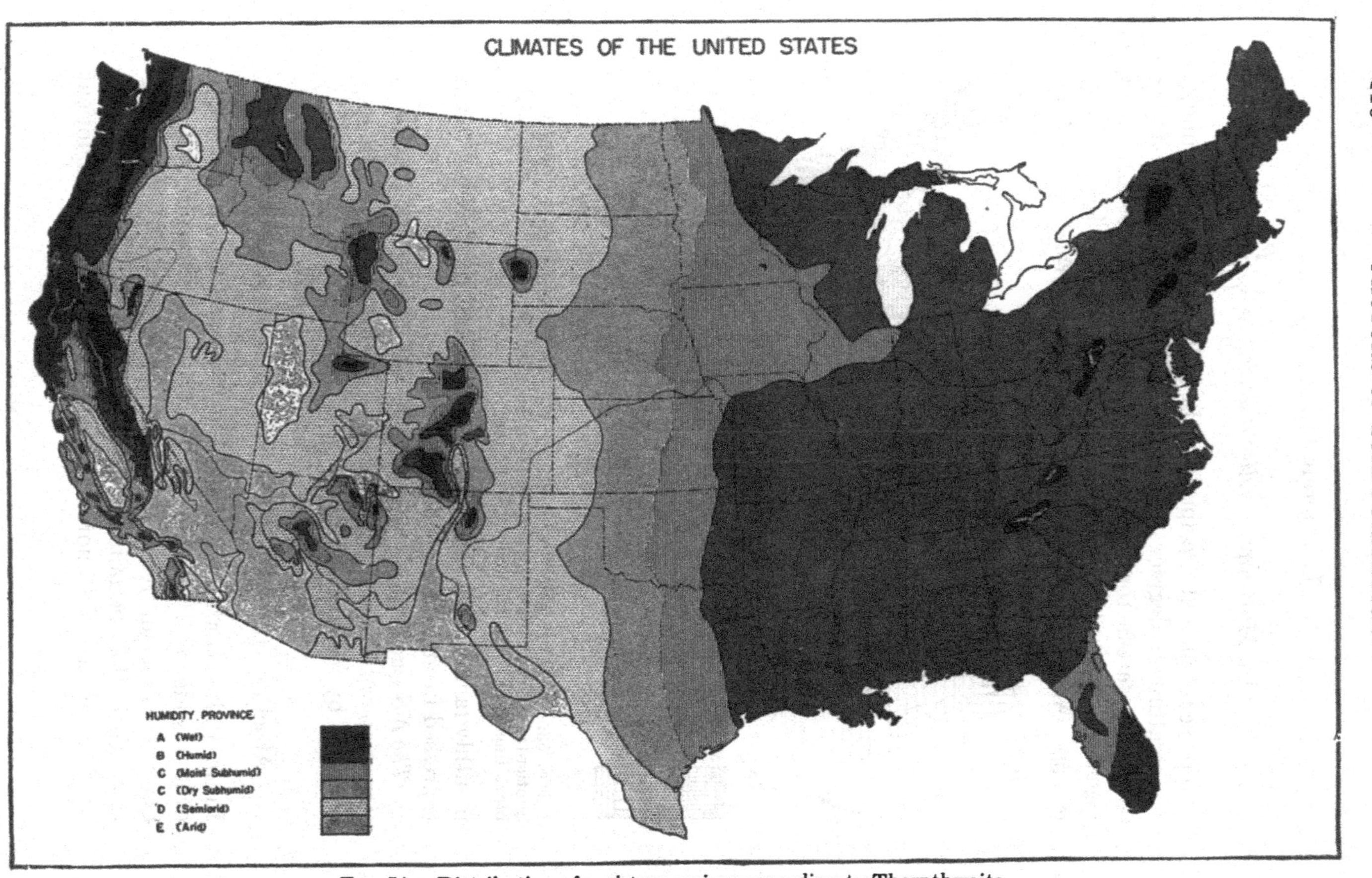

FIG. 54.—Distribution of moisture regions according to Thornthwaite.

maps have been published for Europe (52), the United States, Australia (58), and India (31).

Precipitation Effectiveness or the PE Index.—Thornthwaite (77) has proposed a moisture classification that is based on a summation of monthly moisture values. This "precipitation effectiveness index" may be calculated with the aid of the formula

$$I = \sum_{n=1}^{n=12} 115 \left(\frac{P}{T - 10} \right)_n^{10/9} \tag{18}$$

where P = monthly precipitation in inches,
T = monthly temperature in degrees Fahrenheit,
n = number of months.

In principle, Thornthwaite's value is a more complicated rain factor. Compared with other moisture indexes, it corresponds somewhat better with the distribution of climatic soil types in the Great Plains area.

Crowther's Percolation Factor.—Crowther (14) measured the amount of water that passed through the Rothamsted lysimeters in England, and, by correlating the percolates with rainfall and temperature, he arrived at the following interesting data: "To maintain constant drainage, a temperature rise of 1°F. must be accompanied by 0.75 in. more rain, or, as expressed in the metric system, an increase of 1°C. in temperature requires 3.40 cm. more precipitation to yield constant percolation."

Limitations of Climatic Moisture Indexes.—Generally speaking, all moisture indexes based on ratios suffer from their hyperbolic nature. Whenever the denominator is small, the ratio becomes excessively great, probably out of proportion to its physical or chemical significance.

Correlations between soil properties and moisture, which are based exclusively on rainfall or humidity quotients without taking topography into consideration, neglect the role of runoff, which may reduce or enhance the effectiveness of precipitation in soil-forming processes. As a rule, runoff increases in an exponential fashion with increasing rainfall, but it is difficult to quote accurate figures having general significance.

Furthermore, it should be recognized that the official rainfall and temperature records deal with the *macroclimate.* They are gross values and can only be used as general descriptions of the climate of large areas. Small areas are characterized by a distinct *microclimate* that may differ considerably from the macroclimate. Lower temperatures in depressions, high air humidity along creeks, and modifications in environment due to different types of vegetational cover may be taken as examples. The ultimate goal of students of climatic soil formation is to operate with the actual soil climate (*cl'* instead of *cl*) with all its complicated moisture and temperature relations. Although considerable data on the soil climate proper have been accumulated for a number of localized areas, the results are as yet too scanty to permit comparative analysis. In recent years, the Soil Conservation Service has initiated successfully the experimental evaluation of microclimates and soil climates on an extensive scale.

Are Annual Values Justified?—Within regions of similar seasonal distribution patterns, functional relationships between soil and climate are usually based on annual moisture and temperature data. Annual averages may be wholly fictitious, which is especially true for localities with pronounced diurnal temperature fluctuations. There is, of course, no inherent obstacle in calculating monthly values of rain factors, *NS* quotient, etc., should it be so desired. Indeed, monthly moisture factors may prove an important tool in future refinements of soil-climate investigations.

If a given soil property were correlated with variations in rainfall, any number of curves might be constructed by choosing annual, seasonal, monthly, or even daily rainfall records. For each criterion a different type of curve is likely to result. Which one of the many curves is then the "true" one? No straightforward answer can be given, since there are several modes of judgment. The curve that gives the closest statistical fit might be preferred or the curve that has the greatest slope or the one that may be described by the simplest mathematical equation. Again, quantitative agreement with some preferred theory might be stressed. Whatever the decision may be, it involves an element of choice, and there exists no single rule that might serve as a guide in the selection of curves.

As long as soil-climate functions are studied within regions of uniform seasonal distribution patterns, the annual moisture values give satisfactory correlations.

2. Relationships between Soil Properties and Moisture Factors

The quantitative study of soil property-moisture functions is based on the equation

$$s = f(m)_{T,o,r,p,t,\ldots} \quad (16)$$

The letter s denotes any soil property, and m stands for any climatic moisture index. The variables indicated as subscripts should be kept constant. In nature, this may be accomplished by careful selection of areas that are relatively uniform in all soil-forming factors, save moisture.

The great North American Plateau, which reaches from the Rocky Mountains to the Appalachian Range, is marked by areas of great uniformity in topography, vegetation, and geological strata on the one hand, and diverse climatic conditions, on the other. Here it is relatively easy to select large areas that have great variations in rainfall but little variation in annual temperature. The extensive loess mantle that covers thousands of square miles yields a parent material of remarkable homogeneity and thus permits exclusion of an additional variable. In these regions, topography is level, undulating, or slightly rolling. By restricting soil inspection to level areas or to the crests of the loess elevations, the factor topography also is kept constant. At first sight, such a selection of soil samples may appear arbitrary and artificial, but it affords the only means of arriving at quantitative solutions of the soil-moisture equation.

a. Organic Constituents of the Soil

Surface Soils.—It has long been known that in certain localities humus is almost absent in the soil, whereas in other districts it is present in excessive amounts. Over a century ago, in 1796, de Saussure, in his "Voyages dans les Alpes" (Vol. V, page 208), expressed the opinion that climatic factors are responsible for the existence of different organic-matter levels in soils; however, it is only within the last decade that the problem has been subjected to rigorous quantitative treatment. The difficulty lies in the fact that the amount of soil humus, or, more generally, soil organic matter, present depends upon all the soil-forming factors. Only by the method of separating and controlling the soil formers can the contribution of the individual factors that affect the distribution of organic matter be determined.

Determinations of soil organic matter are usually based on organic carbon analyses, the results of which are multiplied by the

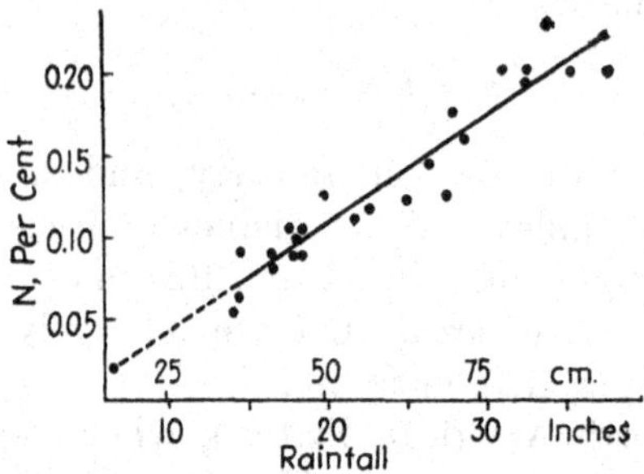

FIG. 55.—Nitrogen-rainfall function for loessial soils to a depth of 10 in. Each point represents the analysis of one soil sample.

conventional factor 1.742. In this book, all humus and organic-matter discussions refer to organic carbon, unless specifically stated. In many soils, the amount of organic carbon is closely related to the total nitrogen content, and under those conditions nitrogen analyses may be taken as an index of organic matter.

TABLE 20.—THE RELATIONSHIP BETWEEN NITROGEN CONTENT OF STABILIZED UPLAND VIRGIN NEBRASKA SOILS AND RAINFALL IN THE TEMPERATE REGION (*Russell and McRuer*)

Precipitation, inches	Soil type	Number of fields sampled	Nitrogen content, per cent	
			0–7 in.	7–12 in.
15.9	Rosebud silt loam	9	0.162	0.101
16.7	Rosebud loam	11	0.145	0.094
19.0	Holdrege silt loam	7	0.160	0.108
23.8	Holdrege silt loam	3	0.209	0.154
28.8	Grundy silt loam	10	0.252	0.172
30.3	Carrington silt loam	22	0.262	0.199
30.4	Marshall silt loam	16	0.252	0.183

NOTE: The average carbon-nitrogen ratio is 11.6 for the surface and 11.1 for the subsoil samples.

Along the annual isotherm of 51.8°F. (11°C.) of the aforementioned loess belt of the Middle West, Jenny and Leonard (39) collected a series of surface samples of virgin and cultivated fields to a depth of 10 in. As may be seen from Fig. 55, a pronounced correlation exists between total soil nitrogen and annual rainfall ($r = +0.946$). The linear relationship may be expressed by the equation

$$N = 0.00655R - 0.023 \tag{19}$$

N indicates the total nitrogen content of soil and R, the mean annual rainfall in inches.

Alway (1) in 1916 and Russell and McRuer (67) in 1927 have reported similar trends for the nitrogen contents of loess soils

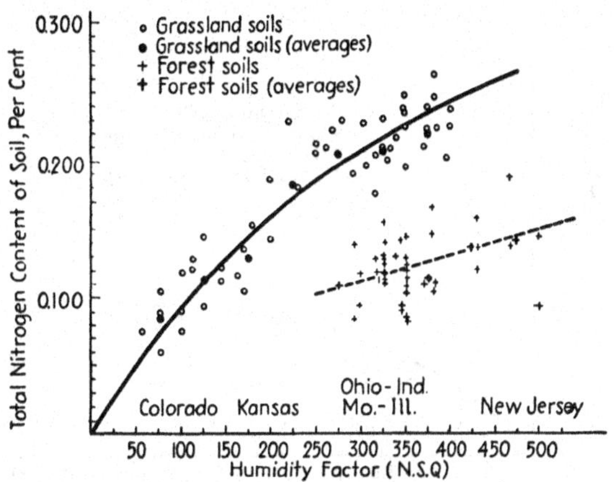

FIG. 56.—Soil nitrogen-rainfall relation along the annual isotherm of 11°C. Black dots represent averages used in calculating Eq. (20).

of Nebraska (Table 20). More recently, Gillam (22) has published organic-matter analyses from the same region.

Jenny (35) has evaluated the nitrogen-moisture function for the region between the Rocky Mountains and the Atlantic Coast. Restricting the study to undulating well-drained upland soils of medium texture, it was found that the nitrogen content of *grassland soils* (mainly cultivated) increases logarithmically with increasing moisture values. The curve shown in Fig. 56 has the mathematical form

$$N = 0.320(1 - e^{-0.0034NSQ}) \tag{20}$$

N denotes the total nitrogen content of the soil to a depth of 7 in., and NSQ indicates Meyer's moisture index.

Data from cultivated soils from originally timbered areas do not show a clear-cut nitrogen-moisture relationship. Moreover, these soils are lower in nitrogen content than the grassland soils from the same climatic regions. The average *carbon-nitrogen ratio* for cultivated grassland soils is nearly constant (from 10 to 12). For this reason, a similar relationship must exist between soil organic matter and rainfall. Indeed, it can be observed by mere field inspection that the grassland soils become darker as one proceeds from the arid to the semihumid regions. This has been shown quantitatively by Gillam (22), who determined the content of black pigment of numerous soils from Colorado and Nebraska.

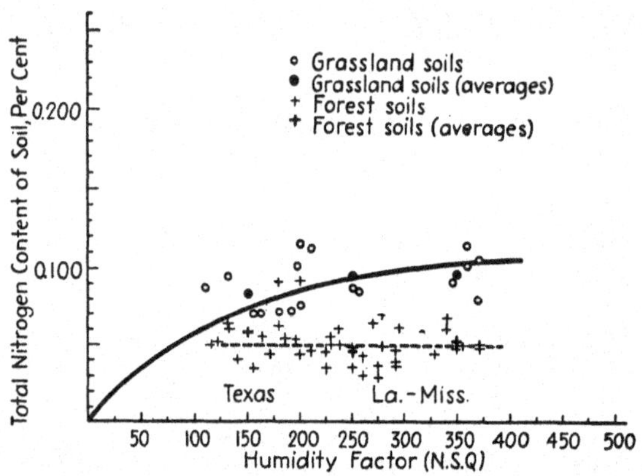

FIG. 57.—Soil nitrogen-rainfall relation along the annual isotherm of 19°C.

In the Southern part of the United States, the nitrogen and organic-matter levels appear to be generally lower than in the Northern region (Fig. 57). The positive trends between nitrogen and *NS* quotient are not so pronounced; in fact, the soils from originally timbered areas give nitrogen values that cluster about a line parallel to the *X*-axis.

Prescott (58) has extended the functional concept to the soils of *Australia* and observed a decided soil nitrogen-*NS* quotient relationship having a correlation coefficient of +0.65.

Generally speaking, the variation of the *carbon-nitrogen ratio* with moisture requires further investigations. For cultivated grassland soils, the carbon-nitrogen ratio appears to be independent of rainfall, whereas, for virgin grassland soils, the quotient has a slight tendency to increase with precipitation.

A similar claim is made by Isaac and Gershill (33) for semiarid and winter rainfall areas of the Cape Province in South Africa.

Profile Functions.—The aforementioned nitrogen-rainfall curves pertain to surface soils to a depth of 7 to 10 in. It is, of course, possible to establish functions for lower soil strata, such as from 10 to 20 in., from 20 to 30 in., or for any depth interval. In such manner, one obtains a family of curves, as shown in Fig.

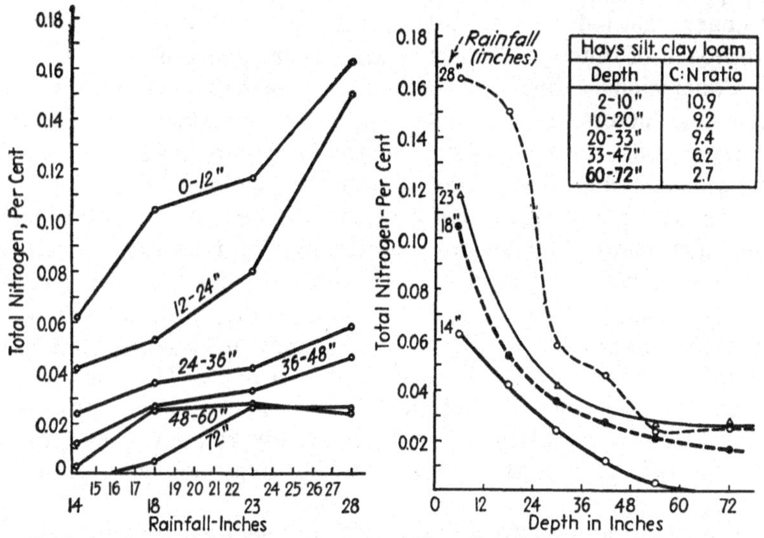

FIG. 58.—Family of curves portraying nitrogen-rainfall functions for a succession of depth intervals (first foot, second foot, etc.).

FIG. 59.—Soil nitrogen-depth functions constructed from the nitrogen-rainfall relations of Fig. 58.

58. It is significant to note that the distances between the curves reveal the profile features of the soils; in fact they may be used to construct depth functions. This is accomplished by selecting a given moisture value and then plotting the corresponding nitrogen contents in relation to depth. An actual example is represented in Fig. 59, the data of which are taken from Fig. 58. The general trend of the nitrogen-depth curve is exponential, the curves for the humid soils tending to lie above those from arid regions. As a rule, the higher the rainfall the deeper the penetration of nitrogen. In similar manner, by combining a series of nitrogen and organic-carbon functions, the vertical distribution of the carbon-nitrogen ratio may be obtained (Fig. 59, inset).

For a number of important soil groups, nitrogen-depth functions have been published by Marbut (48).

The procedure of deriving depth functions from moisture functions may be reversed. With the aid of a series of soil-profile data, it is also possible to construct a family of nitrogen-rainfall functions. This interdependency of the two types of functional relationships may be generalized to include any soil property and any soil-forming factor. In other words, the customary method of soil description by profiles and the new approach of functional analysis supplement one another.

Conclusions.—Under conditions of reasonably constant soil-forming factors, the nitrogen and organic-matter content of surface soils becomes higher as the moisture increases. The relationship is especially pronounced for grassland soils.

Regarding the explanation of the marked effect of moisture on nitrogen, many textbooks express the opinion that the generally low nitrogen content of soils of arid regions is due to a very rapid microbiological decomposition of plant residues, stimulated by neutral soil reaction. However, in view of the vegetation-climate relationships to be discussed in Chap. VII, it is more likely that the scarcity of vegetation, as compared with its abundance under higher rainfall, is mainly responsible for the low nitrogen level of the soils of the western section of the Great Plains.

It might be well to point out that the foregoing nitrogen functions do not necessarily indicate direct causal relationships. In accordance with our definition of soil-forming factors, we are merely attempting to find out how certain soil properties such as nitrogen and organic matter vary with moisture indexes. We are not attempting to elucidate processes of soil formation. For such an undertaking, it would be necessary to analyze the complex interaction between precipitation, runoff, evaporation, transpiration, available growth-water, and a host of related features. This is beyond the scope of the present approach.

b. Inorganic Constituents of the Soil

In *arid regions*, all rain water that penetrates into the soil is either held by the soil particles or moves upward again through evaporation and transpiration by plants. The products of weathering processes are not removed from the soil through leaching. In *humid regions*, a reverse condition predominates. A large part of the water added to the soil percolates through the profile and by way of deep seepage and ground water finally reaches the rivers and oceans. Materials dissolved are leached out. On the basis of these broad principles, it is to be expected that, in general, soils of arid regions are richer in soluble constituents and plant food than those of humid zones. This contention is supported by abundant observational evidence.

TABLE 21.—ANALYSES OF SOILS FROM ARID AND HUMID REGIONS (*Hilgard*)
(Five-day hydrochloric acid digestion, specific gravity 1.115)

Region	Number of analyses	Total soluble material, per cent	Soluble SiO_2, per cent	Al_2O_3, per cent	Fe_2O_3, per cent	CaO, per cent	MgO, per cent	K_2O, per cent	Na_2O, per cent
Arid.......	573	30.84	6.71	7.21	5.47	1.43	1.27	0.67	0.35
Humid.....	696	15.83	4.04	3.66	3.88	0.13	0.29	0.21	0.14

TABLE 22.—AVERAGE CHEMICAL COMPOSITION OF THE SOILS OF THE ARID, PRAIRIE, AND HUMID REGION (*Coffey*)
(Acid-digestion method)

Region	Number of samples	CaO, per cent	MgO, per cent	K_2O, per cent	P_2O_5 per cent
Arid....................	318	2.65	1.20	0.71	0.21
Transition (prairie)......	215	1.09	0.51	0.43	0.18
Humid (forest)..........	743	0.41	0.37	0.37	0.16

Hilgard's and Coffey's Data.—Hilgard (28) has analyzed a great number of soils from the arid and the humid regions of the United States. Table 21 shows the results for soils not derived from or underlaid by limestone formations. Hydrochloric acid dissolves more total material from the arid than from the humid soils. In addition, the arid soils are higher in content of calcium, magnesium, potassium, and sodium, both in absolute amounts and in relation to alumina. These data are most easily explained on the assumption that the higher rainfall of the humid regions impoverishes the surface soil through leaching. Hilgard's analyses served as a cornerstone in the earlier development of the concept of climatic soil types.

Coffey (10) reports a comparison of soils from different moisture regions based on chemical analyses made in the United States by the acid-digestion method from 1891 to 1909. The absolute percentage figures in Table 22 depart considerably from those quoted by Hilgard, but they agree in showing much larger percentage figures in the arid than in the humid soils. The soils of the transition zone occupy an intermediate position, a fact already observed by Hilgard.

Leaching Values of Potassium and Sodium.—Hilgard's contention that high rainfall impoverishes the soil through leaching is convincing by virtue of the great number of analyses, which eliminates chance correlation. More accurate insight is gained by comparing the chemical composition of the soil with that of the parent rock from which the soil originated. Potassium (K) and sodium (Na) are particularly sensitive criteria of leaching intensities, and their relative rate of translocation can be measured by comparing the ratio $\frac{K_2O + Na_2O}{Al_2O_3}$ of the most-leached horizon with that of the parent material (see page 27). The quotient formed by the two ratios has been called the leaching value β. The smaller β the more pronounced is the relative leaching of the elements as a result of weathering and soil formation.

For soils derived from similar parent rock such as "sedimentary material containing carbonates," which includes moraines, loess, shales, and sands, the influence of rainfall on the translocation of

TABLE 23.—LEACHING OF K + NA AS INFLUENCED BY MOISTURE VARIATIONS WITHIN THE TEMPERATE AND COLD ZONES [*Jenny* (36)]
(Fusion analyses of soils)

Region	Number of profiles	Leaching value β
Semiarid to semihumid (chestnut- and chernozemlike soils)	15	0.981 ± 0.059
Semihumid, North Dakota (chernozemlike soils)	29	0.901 ± 0.028
Humid (podsolized soils)	12	0.719 ± 0.053

K and Na can be clearly detected (Table 23). Within the temperate zone, β becomes smaller as the moisture values increase.

In other words, the relative migration and leaching of potassium and sodium in the soil are definitely more pronounced in humid than in arid regions.

Calcium-rainfall Functions.—The analyses of Nebraskan loess by Alway (1) and coworkers may be used for functional studies on removal of CaO by percolating rain water.

Soils from 30 virgin prairie fields were collected and composite samples from 50 borings completely analyzed. Table 24 clearly emphasizes the pronounced negative correlation between annual precipitation and HCl-soluble CaO as well as HCl-insoluble CaO.

The decline of CaO may be expressed mathematically by the equations

$$\text{Acid-soluble CaO} = 15.24e^{-0.040P} \quad (21)$$

$$\text{Acid-insoluble CaO} = 1.98e^{-0.024P} \quad (22)$$

where P = annual precipitation from 40 to 80 cm.,
e = base of natural logarithms.

Significant differences exist in the *rate* of removal of CaO of the two groups of compounds

$$\text{For acid-soluble CaO:} \quad \frac{d(\text{CaO})}{dP} = -0.040 \cdot (\text{CaO}) \quad (23)$$

$$\text{For acid-insoluble CaO:} \quad \frac{d(\text{CaO})}{dP} = -0.024 \cdot (\text{CaO}) \quad (24)$$

TABLE 24.—AMOUNT OF CaO IN NEBRASKAN LOESS PROFILES TO A DEPTH OF SIX FEET (*Alway*)
(Annual temperature 49.4 to 51.8°F.)

Locality	Annual precipitation, inches	HCl-soluble CaO		HCl-insoluble CaO	
		Analyses (Alway), per cent	Calculation,* per cent	Analyses (Alway), per cent	Calculation,* per cent
Wauneta	18.55	2.26	2.32	0.72	0.64
McCook	19.08	2.42	2.19	0.53	0.62
Holdrege	24.24	1.32	1.30	0.47	0.45
Hastings	26.87	1.05	1.00	0.45	0.39
Lincoln	27.51	0.83	0.93	0.31	0.37
Weeping Water	30.19	0.78	0.71	0.31	0.31

* Based on Eqs. (21) and (22).

According to these differential coefficients, the rate of leaching of the acid-soluble CaO (mainly from $CaCO_3$ and exchangeable

Ca) is greater than that of the insoluble form (mainly from unweathered Ca-alumino silicates).

By extrapolating Eqs. (21) and (22) to $P = 0$, one may calculate the total CaO content of theoretically unweathered loess, which amounts to 17.2 per cent CaO, a value that is in agreement with analyses of unaltered loess reported in the litera-

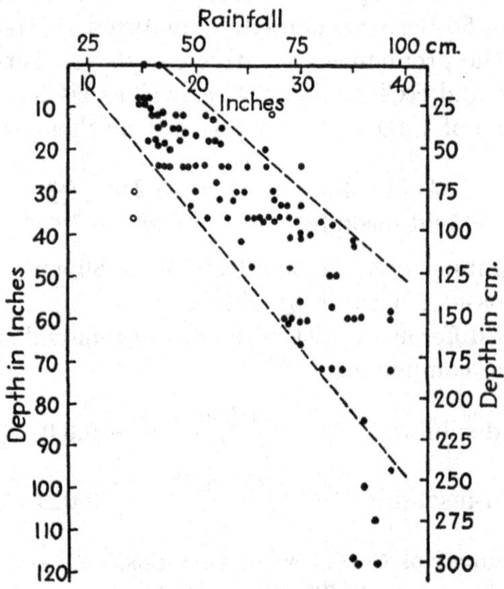

FIG. 60.—Relation between depth of carbonate accumulation and rainfall in loessial soils. Every point represents the depth of the beginning of the concretion zone in a soil cut investigated.

ture. Extrapolating Eq. (21) to the rainfall of humid regions yields a content in acid-soluble CaO of less than 0.50 per cent, which checks with both Hilgard's and Coffey's averages in Tables 21 and 22.

The Carbonate Horizon and the Pedocals.—Marbut (47) has called attention to the carbonate horizon, which is an outstanding visible soil characteristic. Well-developed soils in regions of low rainfall have in their profile a layer of carbonate concretions that contains more $CaCO_3$ and $MgCO_3$ than the horizons above or below (Table 25). In well-drained and highly developed soils of humid regions, the carbonate horizon usually is missing.

The position of the upper part of the lime carbonate horizon has been carefully recorded along a transect extending from the semiarid region of Colorado through Kansas to the humid areas of Missouri (39). In Fig. 60, every point represents the observed depth of the beginning of the lime concretions in a profile. The

TABLE 25.—CHEMICAL EVIDENCE OF THE CARBONATE HORIZON IN TWO SOILS FROM REGIONS OF LOW RAINFALL

Soil from Stalingrad U.S.S.R. (23)		Soil from Krydor, Saskatchewan, Canada (49)			
Depth, inches	CO_2 content, per cent	Depth, inches	CaO, per cent	MgO, per cent	CO_2 from carbonates, per cent
2.0– 3.9	0.063	0– 5	1.70	1.00	0.00
9.8–11.0	0.048	8–14	2.27	1.38	1.06
15.7–17.7	0.097	14–24	**8.40**	**3.13**	**7.72**
19.7–21.7	3.186	24–65	1.97	2.19	0.85
23.6–25.6	4.375				
29.5–31.5	5.890				
33.5–35.4	**7.136**				
39.4–40.4	6.230				
40.4–43.3	5.430				
49.3–51.2	4.321				
59.1	3.165				

Comparison of the CaO and MgO figures indicates that the carbonate layer is predominantly a lime horizon. (CaO and MgO figures are based on fusion analyses.)

TABLE 26.—DEPTH OF LIME HORIZON IN NEBRASKAN LOESS SOILS (66)
(Annual temperature from 48 to 50°F.)

Approximate Annual Precipitation, Inches	Depth of Carbonate Horizon, Inches
20	12–24
25	35–47
30	About 60

corresponding rainfall data were interpolated from surrounding meteorological stations. Although great variations in depth exist, a general increase of depth with rainfall is obvious. For the rainfall interval, from 12 to 40 in., the thickness of the surface layer, which is free of carbonates, augments, on the average, 2.5 in. for each additional inch of precipitation. Somewhat similar conditions appear to exist in the loess region of Nebraska, as shown by data from Russell and Engle (Table 26).

Marbut has chosen the carbonate horizon as a fundamental criterion for the classification of soils. All soils that possess an accumulation horizon of lime are designated by Marbut as *pedocals*.

CaO-depth Functions of the Great Climatic and Vegetational Soil Groups (Zonal Soils).—The number of calcium-rainfall functions is too limited to permit the derivation of calcium-depth functions. It is, however, possible to get an approximate idea of the type of CaO-depth distributions in relation to rainfall by

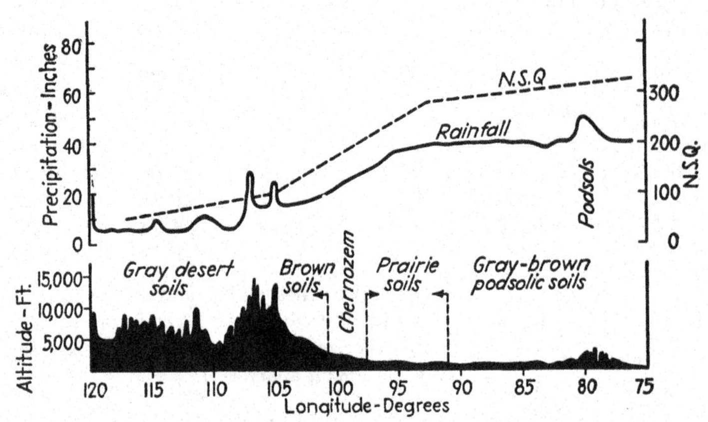

FIG. 61.—Relief, rainfall, *NS* quotients, and sequence of zonal soil types approximately along the 11°C. isotherm in the United States.

utilizing profile analyses of the great climatic soil groups. Marbut, who successfully applied the Russian ideas of soil classification to the United States, distinguished between the following broad soil groups in the temperate zone (compare Figs. 61 and 96)·

Zonal Soil Groups	Approximate Annual Rainfall Limits along the 11°C. Isotherm, Inches
Gray desert soils	<15
Brown soils (arid brown and chestnut soils)	15–20
Chernozems	20–30
Prairie soils	30–40
Gray-brown-podsolic soils	30–50

These soil groups possess characteristic profile patterns of soil properties. In the following paragraphs, the important CaO-depth functions will be briefly examined. Inasmuch as the soil-forming factors are not rigidly controlled, variations in parent material, topography, etc., will bring about considerable irregularities within the various soil groups.

The curves assembled in Figs. 62 to 66 were constructed from chemical data contained in Marbut's Soils of the United States. CaO percentages of dry soils are plotted on the ordinates; depths of the profiles are indicated on the abscissas. The numbers listed with the state designations refer to the tables of analyses in Marbut's Atlas (49), and thereby identify the soil types

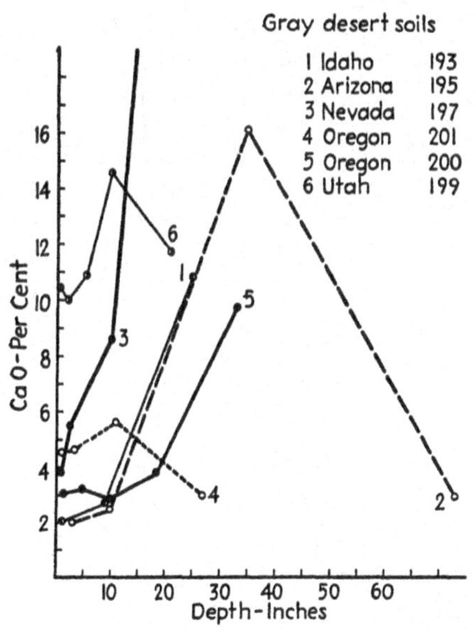

FIG. 62.—Relationship between CaO content and depth for gray desert soils.

selected for the compilation. Only soils derived from calcareous sedimentary materials are included in the graphs.

Gray Desert Soils.—The general pattern has an erratic appearance, though in all cases CaO is higher in the subsoil than in the surface soil. The enormous variations in the lime content of the surface soils probably are associated with the heterogeneity of the parent materials. Lime horizons show no consistent position as regards depth.

Brown Soils.—The arid brown soils and the chestnut soils are united in this group. Nearly all the profiles were collected by Marbut, who, unfortunately, did not sample the horizons below the carbonate zone. Again there is considerable variety in the distribution patterns, but, unlike the desert profiles, the surface horizons (from 0 to 5 in.) of the brown soils never exceed 2 per cent CaO. A significant downward movement of carbonates, as a result of rainfall, is clearly demonstrated. Most of the lime horizons occur at a depth of from 10 to 30 in.

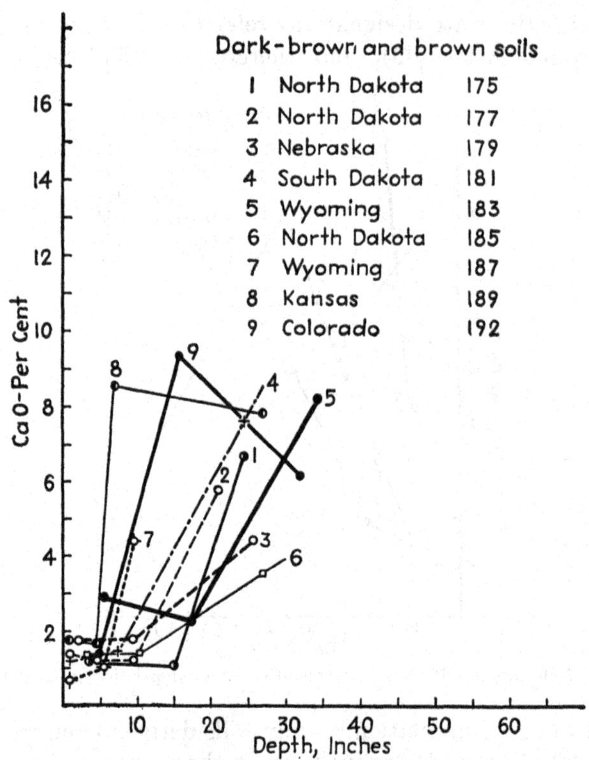

FIG. 63.—Relationship between CaO content and depth for dark-brown soils (chestnut soils) and brown soils.

Chernozems.—The general features are similar to those of the brown soils, yet the effects of higher rainfall are well manifested. The surface layers with less than 2 per cent CaO have an average thickness of 10 in. or more, *i.e.*, twice that of the brown soils. The depth of the accumulation zone, though variable, fluctuates between 20 and 40 in. In harmony with the pronounced removal of CaO from the surface, there is a high absolute concentration of calcium in the accumulation zone.

Prairie Soils.—This graph differs radically from the preceding ones, perhaps more so than one would anticipate from moisture considerations. The CaO content of less than 1.5 per cent in the surface horizons is perhaps to be expected, but the absence of carbonates, even at great depth, for the majority of the profiles

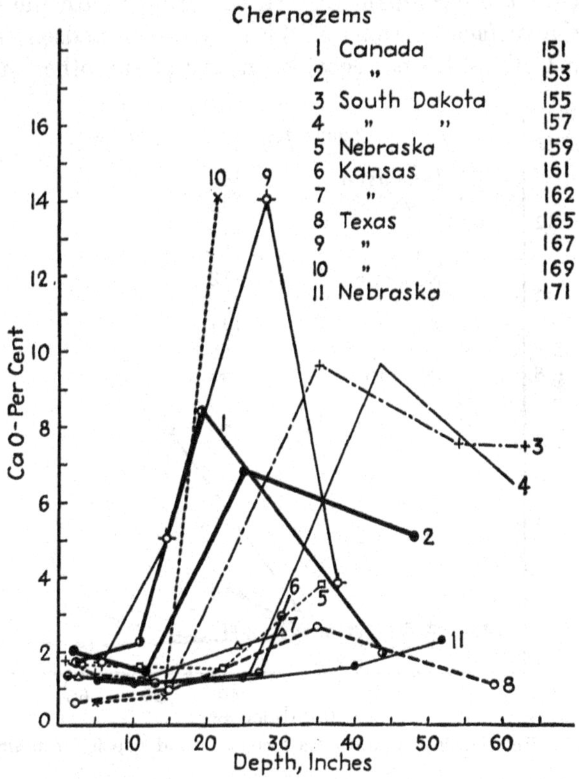

FIG. 64.—Relationship between CaO content and depth for chernozems.

is not easily accounted for. In comparison with the chernozems and the gray-brown-podsolic soils, the prairie soils appear to be in a class by themselves.

Gray-brown-podsolic Soils.—Only soils derived from calcareous sedimentary deposits are included in Fig. 66. Compared with the prairie soils the general pattern of the forest soils is surprising, because of the high CaO content at moderate depths. Presumably, the steep curves are not lime horizons but represent the calcareous parent material (*C* horizons). Inasmuch as the rainfall in the timbered regions is somewhat higher than in the prairie belt and leaching under forest cover is more pronounced than under grass (page 225), a deeper *solum* than is evident from the graphs would be expected. Differences in permeability of the parent material and in its lime content possibly would account for the discrepancies. In conformity with the environment, the surface horizons of the gray-brown-podsolic soils are lower in CaO (<1.0 per cent) than any of the other major soil

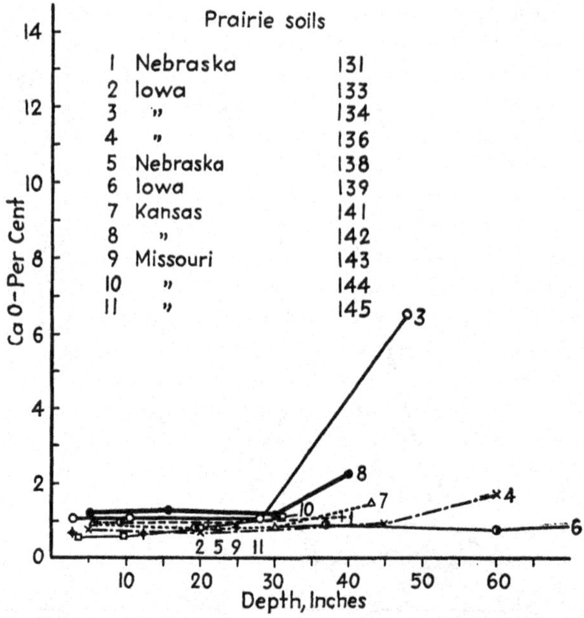

FIG. 65.—Relationship between CaO content and depth for prairie soils.

groups of the temperate region. Generally speaking, the CaO-depth functions of the forest soils seem to constitute logically a moisture sequence to the diagrams of the chernozems rather than to those of the prairie group.

The Iron Horizon and the Pedalfers.—In contrast to the carbonate horizon of the pedocals, well-drained and mature soils in humid regions are characterized by an iron zone that consists largely of iron hydroxide in various stages of dehydration. In many soils of northern countries, the iron horizon is present in the form of a brown to dark-brown layer, often cemented by gel-like iron compounds. In subtropical and tropical soils, iron-oxide concretions frequently reach the size of walnuts and occasionally assume even greater dimensions.

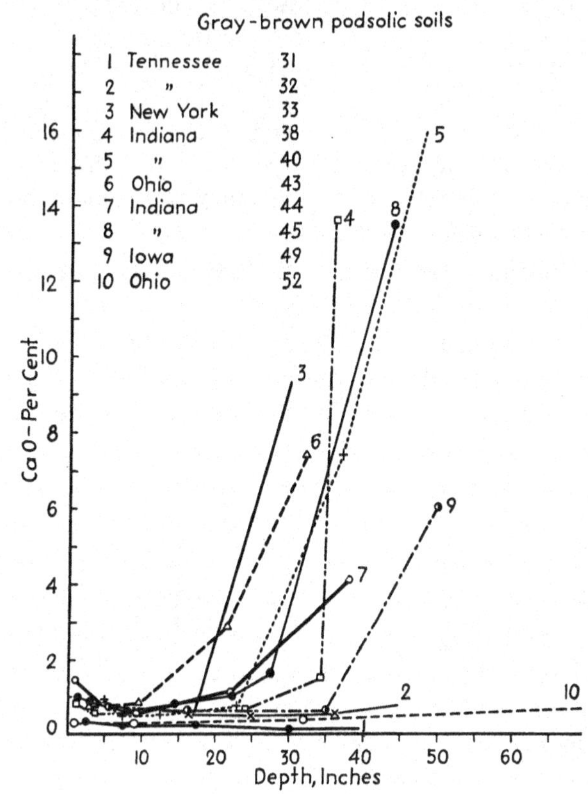

Fig. 66.—Relationship between CaO content and depth for gray-brown podsolic soils.

In chemical descriptions of profiles, the presence of the iron horizon is indicated by an increase in the Fe_2O_3 content at some depth in the soil (Table 27). The formation of an iron horizon is not restricted to any particular type of parent material; it is a distinguishing property of well-aged soils of humid regions. Marbut classifies all soils with a typical iron horizon as *pedalfers*.

Properties of Colloidal Clays.—During the process of weathering, the silicon and aluminum tetrahedra and octahedra, which are the building stones of the alumino silicates of igneous parent rocks, rearrange themselves and form colloidal clay particles.

Most colloid chemists (38) restrict the term "colloidal" to particles that have diameters from 1 to 100 millimicrons (mμ), but soil scientists set the upper limit as high as 1,000, 2,000, or even 5,000mμ (= 5μ = 0.005 mm.). Colloidal clay particles are of vast importance, because they control to a great extent the physical and chemical behavior of soils. Water permeability, aeration, horizon development, swelling and shrinkage, and the development of soil structure depend on the amount and kind of colloidal clay in the soil. Likewise, the growth of plants is affected by the soil colloids, because they are the storehouse for many important nutrient elements.

TABLE 27.—CHEMICAL INDICATION OF THE IRON HORIZON IN A SOIL OF THE HUMID REGION
(Fusion analysis)

Heath podsol on sand (87) (Denmark)		
Depth, inches	SiO_2, per cent	Fe_2O_3, per cent
0– 2.7	75.98	0.72
2.7– 5.9	94.85	0.98
5.9– 8.7	62.69	**5.20**
8.7–11.4	90.30	2.26
29.5–33.5	94.14	0.58

Clay Content of Soil and Rainfall.—The enhancement of weathering in regions of higher rainfall is well illustrated by the mineralogical analyses of soils conducted by the former Bureau of Soils. For purposes of comparison, the minerals identified in the sand and silt portions were divided into two groups:

a. Quartz,

b. Minerals other than quartz.

Coffey (10) has arranged the data according to three main moisture groups (Table 28), and he demonstrates conclusively a preponderance of quartz in soils of high rainfall. For example, out of 100 mineral particles from the sand portion of arid soils, only 63 are quartz; whereas, in the case of humid soils, 92 are quartz. Since quartz is a mineral that weathers very slowly, its relative accumulation must be due to the disappearance of other minerals.

Since the inauguration of the U. S. Soil Survey some 40 years ago, thousands of texture analyses of American soils have been published in the Soil Survey Reports. Although the technique of mechanical analysis has been materially improved during the

TABLE 28.—AVERAGE PERCENTAGES OF MINERALS OTHER THAN QUARTZ IN THE SOILS OF THE ARID, TRANSITION, AND HUMID REGIONS

Region	Number of soils examined	Minerals other than quartz	
		Sand portion, per cent	Silt portion, per cent
Arid	30	37	39
Transition (prairie)	40	20	29
Humid (forest)	160	8	12

last decade, the older data nevertheless are suitable for quantitative studies of the relationships between soil and climate (39). In the temperate region, the clay content (particles $<5\mu$) of the soil is positively correlated with annual rainfall. The scatter

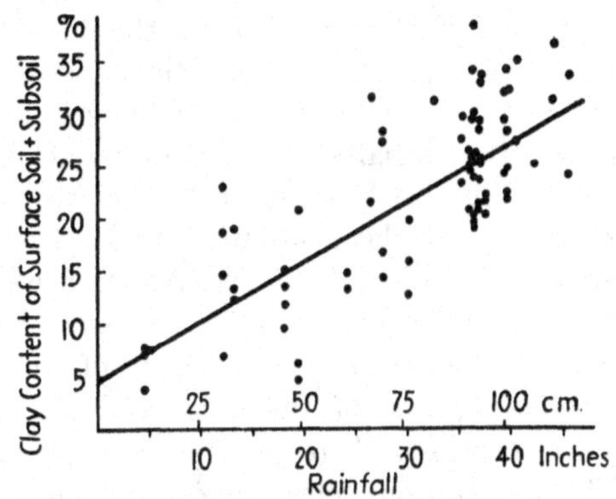

FIG. 67.—Average clay content to a depth of 40 in. of soils derived from various parent materials. Every point represents one profile. Mean annual temperature 52 to 56°F.

diagram shown in Fig. 67 has a significant correlation coefficient of +0.76. Applying the method of least squares, the general trend may be expressed by the equation

$$C_1 = 0.567P + 4.52 \tag{25}$$

C_1 indicates the average clay content of a soil profile to a depth of approximately 40 in., and P denotes the annual rainfall in inches. The rather wide variability of the individual profiles is probably due to variations in parent material that was not kept constant. Different types of sedimentary rocks such as loess, moraines, limestones, and lacustrine deposits were included in the series.

A more significant correlation is obtained within the loess belt of Kansas and Missouri, as is brought to light in Fig. 68. Here the content of colloidal clay ($<2\mu$) of the surface soil to a depth of 10 in. gives a correlation coefficient of +0.82 with rainfall. The equation has the form

$$C_0 = 0.914P + 1.33 \quad (26)$$

in which C_0 refers to 2μ clay.

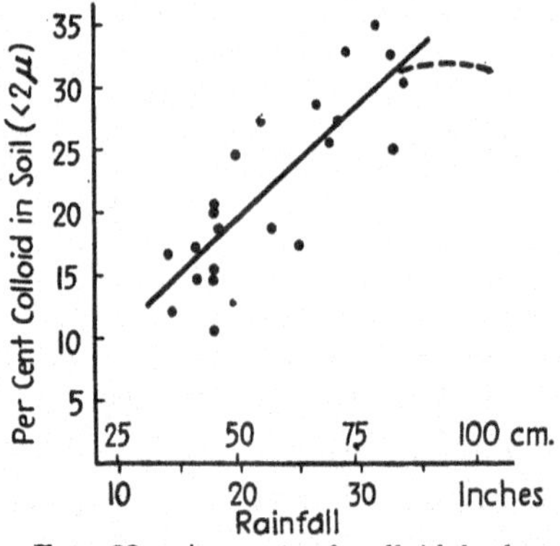

Fig. 68.—Amount of colloidal clay of the surface layer (from 0 to 10 in.) of soils along the 11°C. isotherm. Each dot represents one soil.

An entirely different type of clay-rainfall relationship was found by Craig and Halais (12, 13) on the tropical island of Mauritius. The mechanical analyses of 54 soils that were derived from doleritic basalts are summarized in Table 29. Increased precipitation clearly reduces the amounts of stone and gravel and also lowers the clay content. This latter observation is in conflict with the findings in North America where the clay content increases with higher rainfall. However, the British investigators make it clear that the lowering of the clay portion and the corresponding rise of the sand and silt portion with rainfall does not correspond to an increase in unweathered fragments. The coarser particles consist instead of secondary concretions that represent a more advanced state of decomposition of the original material than does the clay fraction. It appears that, in the humid tropics, pellets rich in sesquioxides accumulate, rather than siliceous colloidal clays. In comparing these relationships with the curves observed in the Midwestern states, it should be noted that in Mauritius not only annual temperature but also annual precipitation reach very high values. Were it possible to find within the temperate belt of North America equally high values of precipitation (from 75 to 150 in.), it would be quite conceivable that the curve in Fig. 68 might reach a maximum and then decline as a consequence of decomposition of clay particles.

TABLE 29.—MECHANICAL COMPOSITION OF MAURITIUS SOILS
(Surface soils)

Mean annual rainfall, inches	Stone and gravels 200–2 mm. in diameter, per cent	Inorganic colloids (clay <2μ), per cent	Sands and silts, per cent
25– 50	8	73	27
50– 75	4	67	33
75–100	5	56	44
100–125	5	49	51
125–150	2	45	55

Exchangeable Cations.—Colloidal clay and humus particles carry cations (Ca, Mg, Na, K, H) that may be replaced by other cations such as NH_4 or Ba. These adsorbed ions are called

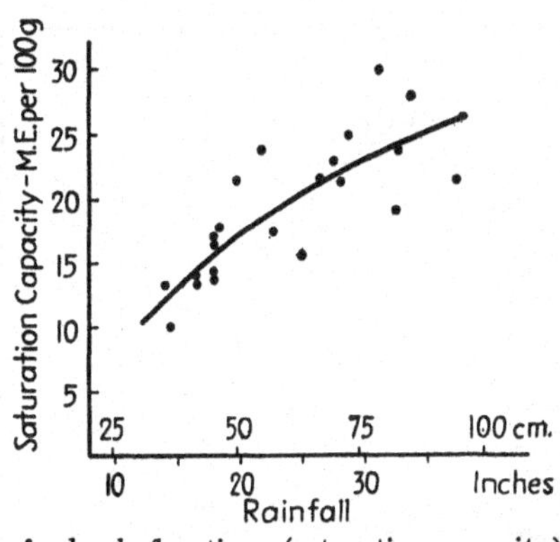

FIG. 69.—Amount of adsorbed cations (saturation capacity) of loessial soils in relation to rainfall.

"replaceable" or "exchangeable" cations. They play a significant role in the nutrition of plants and, to a great extent, affect many physical properties of soils (38).

The total amount of exchangeable cations determined by leaching the soil with a replacing reagent of an arbitrarily fixed pH value (usually pH 7) is known as the base exchange or saturation capacity of the soil. It is generally expressed as milliequivalents of cations per 100 g. of oven-dry soil.

Jenny and Leonard determined the saturation capacity of the loessial soils described on page 114. The values, plotted as a function of rainfall, are given in Fig. 69. The saturation capacity increases logarithmically with rainfall ($r = +0.815$). Moreover,

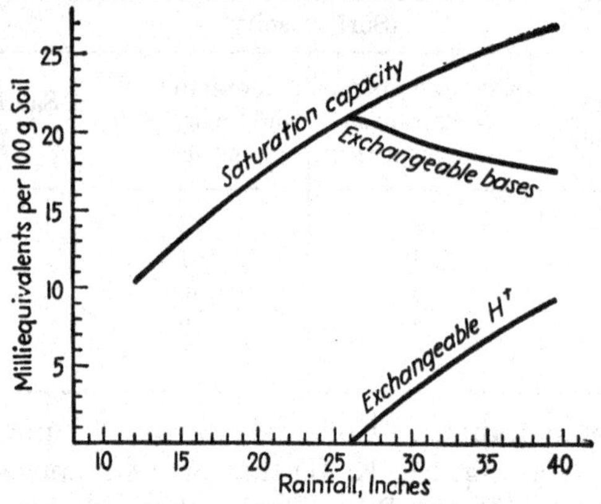

FIG. 70.—Showing saturation capacity, exchangeable bases, and exchangeable hydrogen ions of the surface layer of loessial soils as a function of rainfall.

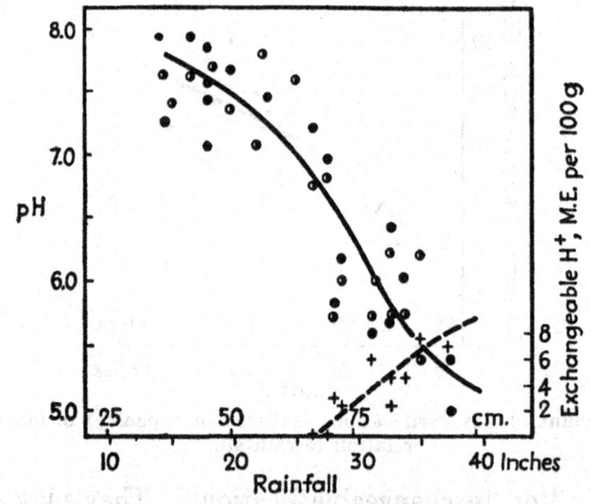

FIG. 71.—Relation between soil acidity and rainfall for surface soils derived from similar parent material (loess). Circles stand for pH, crosses for exchangeable hydrogen ions.

they examined the nature of the adsorbed cations as a function of moisture. If we distinguish between hydrogen ions (H) and all other replaceable cations (Ca, Mg, K, Na, etc.), we have the following relationship:

Saturation capacity = Exchangeable H + Total exchangeable bases

The change of these two components of the saturation capacity with rainfall is presented graphically in Fig. 70. The sum of the exchangeable bases in the arid region is low and equal to the saturation capacity. With the leaching reagent adopted (neutral Ba-acetate solution), no exchangeable hydrogen ions are found in these soils. The combined bases increase with moisture up to

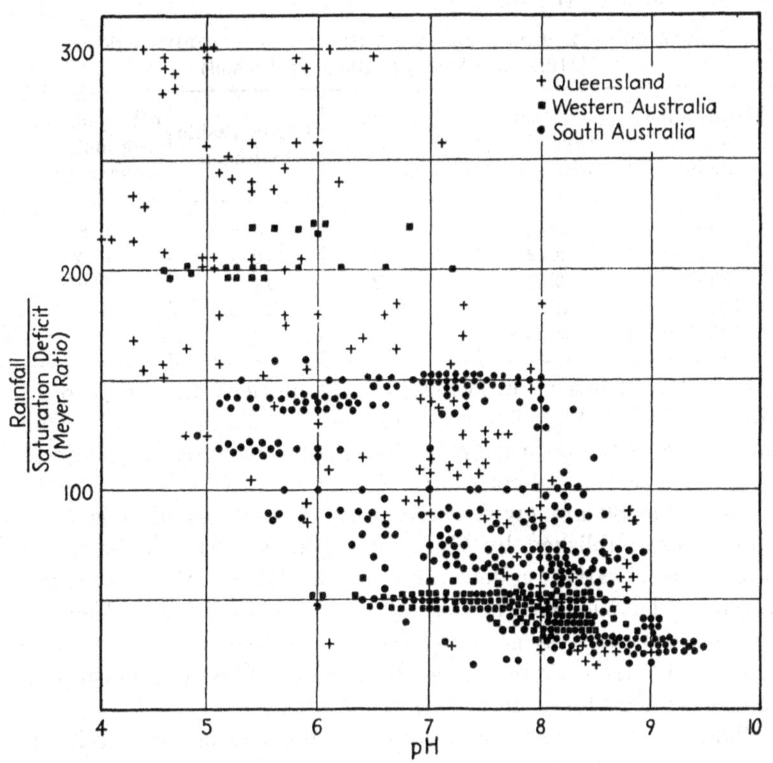

FIG. 72.—Relationship between *NS* quotient values and pH of Australian surface soils. (*Prescott.*)

a rainfall value of 26 in., or *NS* quotient = 202. At this point a significant change takes place. Whereas the saturation capacity continues to rise with moisture above this point, the exchangeable bases begin to decline. Simultaneously hydrogen ions appear, in compensation for the reduction of bases with respect to the saturation capacity. The maximum of the exchangeable base curve corresponds to the chernozem belt.

The increase of exchangeable hydrogen ions is reflected in the pH values of the soils (Fig. 71). Under low rainfall, pH values are high, denoting alkalinity. Under high rainfall, pH values are low, indicating soil acidity. The neutral point (pH = 7) occurs in the chernozem belt. Notwithstanding the marked variability of the pH values, the general trend of soil reaction with rainfall clearly comes to the fore.

TABLE 30.—EXCHANGEABLE BASE STATUS OF MAURITIUS SOILS
(Milliequivalents per 100 g. of dry soil)

Mean annual rainfall, inches	Saturation capacity (pH = 7)	Total exchangeable bases	Exchangeable hydrogen	pH, collodion bag method, colorimetric
25– 50	29.5	24.0	5.5	6.8
50– 75	26.2	15.9	10.3	6.3
75–100	22.9	8.2	14.7	5.95
100–125	22.3	5.4	16.9	5.7
125–150	20.6	4.0	16.6	5.6

Prescott (58) has published a pH-*NS* quotient scatter diagram for Australia (Fig. 72) that exhibits a similar trend. The scattering of the data is great, because no attempt was made to keep parent material constant.

In Table 30 are given corresponding analyses of soils from Mauritius, published by Craig and Halais (13). In harmony with the negative colloid-rainfall correlation, the saturation capacity declines with increasing moisture conditions. Increasing precipitation is accompanied by decreasing exchangeable bases and rising replaceable hydrogen. This relationship is similar to that found in the United States.

The formation of soil acidity under conditions of high precipitation may be schematically represented as follows:

$$\underbrace{\boxed{\text{Colloid}}\begin{matrix}\text{Ca}\\ \text{Mg}\\ \text{K}\end{matrix}}_{\text{Neutral clay}} + \underbrace{\begin{matrix}\text{HOH}\\ \text{HHCO}_3\end{matrix}}_{\text{Water and carbonic acid}} \rightarrow \underbrace{\boxed{\text{Colloid}}\begin{matrix}\text{H}\\ \text{H}\\ \text{H}\\ \text{H}\\ \text{H}\\ \text{H}\end{matrix}}_{\text{Acid clay}} + \underbrace{\begin{matrix}\text{KOH}\\ \text{Ca(HCO}_3)_2\\ \text{Mg(HCO}_3)_2\end{matrix}}_{\text{Removed from surface horizons by leaching}} \quad (27)$$

The exchangeable bases of the neutral clays are replaced by hydrogen ions of water and carbonic acid. This interchange of ions converts the neutral clay into a hydrogen clay or acid clay. The hydroxide and bicarbonates formed as a result of the reaction are leached out by percolating rain water, whereas the acid clay remains in the soil and gradually accumulates. Increasing amounts of rainfall progressively intensify these processes.

To summarize the broader relationships between the exchangeable cations held on clay particles and the climatic moisture regions in Northern United States, the following schematic presentation may prove helpful

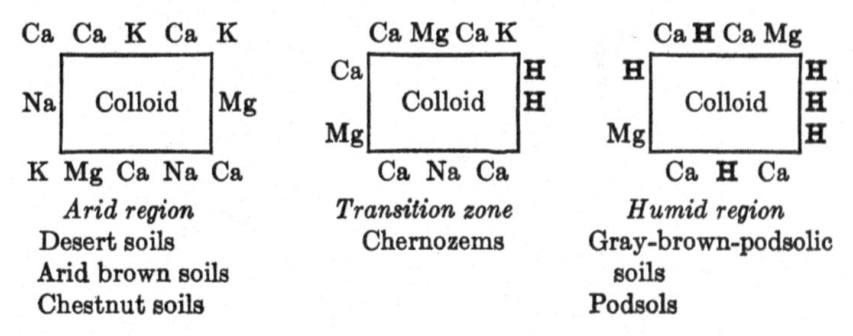

Soils of arid regions are generally low in colloids and consequently possess low saturation capacities. The exchangeable cations are chiefly Ca, Mg, K, and Na. Depending on local ground-water conditions, Mg and Na ions may predominate (alkali soils). In the more humid parts of the chernozem belt, leaching assumes

TABLE 31.—NATURE OF ADSORBED IONS OF SURFACE SOILS [*Gedroiz* (21), *Kelley* (41), *and de Sigmond* (75)]

Moisture region	Percentage composition of adsorbed cations					
	Na	K	Mg	Ca	H	Total
Arid (alkali soil).............	30	15	20	35	0	100
Transition (chernozem).......	2	7	14	73	4	100
Humid (podsol)..............	Trace	3	10	20	67	100

significant proportions, and hydrogen ions begin to appear on the colloidal particles. Saturation capacities are high because the soils are rich in clay and humus colloids. Under still higher precipitation, the downward percolation of rain water becomes a dominant feature. The bases are for the most part leached out, and hydrogen ions constitute the bulk of the adsorbed cations.

Table 31 shows the relative proportions of the various exchangeable cations in soils of the great moisture regions of the cooler climates.

Constitution of Soil Colloids.—Although the exchangeable cations play a dominant role in soil behavior, they comprise but a small fraction (usually less than 5 per cent) of the total mass of the soil colloids. Depending on soil-forming conditions, the major portion of soil colloids consists of alumino silicates (clay) and a variety of organic materials (humus).

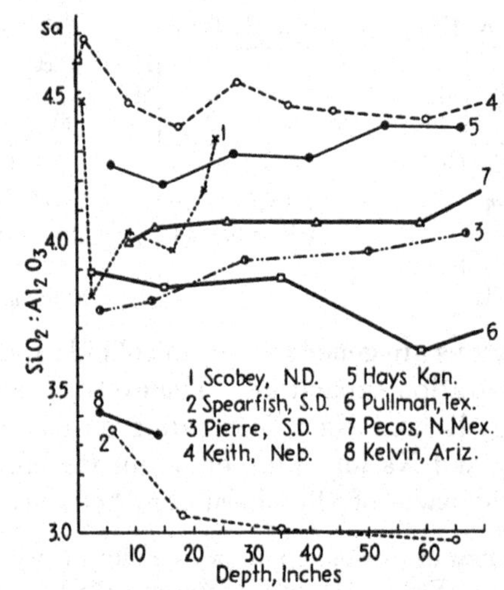

FIG. 73.—Silica-alumina ratios of colloids extracted from soils of arid and semi arid regions.

First, let us examine a number of silica-alumina ratios of soil colloids from arid regions in their relation to depth by making use of data contained in a publication by Brown and Byers (6) on the chemical and physical properties of dry-land soils and of their colloids. As seen from Fig. 73 all ratios fall into the range from 2.9 to 4.7 which, compared with ratios from soils of humid tropical regions, must be considered as high. The spread of the values is due mainly to the composition of the colloids of the parent materials, which are, with few exceptions, sedimentary deposits. The trend of the depth functions is not consistent. In general, however, soil formation in arid regions has brought about a relatively small deviation from the original composition of the colloidal particles.

The effect of increasing rainfall on the *sa* value of colloids is not yet well understood. In cool regions, the clay fraction of surface soils appears to have a tendency to enrich in silica as precipitation becomes higher, but the available data are too scanty to reach a definite conclusion. In warm regions, the variation in clay composition with rainfall is more conspicuous (62). High amounts of precipitation tend to produce colloids with low silica-alumina ratios.

The aforementioned doleritic basalts of Mauritius, which are very nearly quartz free, have a silica-alumina ratio of approximately 5.8. The corresponding value for mature soils is 1.87 at low rainfall and 0.43 at high rainfall. During weathering and soil formation, the silica molecules must have been removed at a faster rate than those of alumina. The ratio of ferric oxide to alumina is 0.77 in the original rock material and from 0.6 to 0.7 in all the soils. For this reason, these two substances may be considered equally permanent fractions.

Fusion analyses of inorganic colloids ($<2\mu$ diameter) of representative soils are listed in Table 32. The *sa* values are very low, particularly in the more humid regions. The high content of water above 110°C. is indicative of crystalline colloidal material (42).

TABLE 32.—CHEMICAL ANALYSIS OF INORGANIC COLLOID ($<2\mu$) FROM MAURITIUS SOILS

Mean annual rainfall, inches	Number of samples analyzed	$\frac{SiO_2}{Al_2O_3}$	$\frac{SiO_2}{Al_2O_3 + Fe_2O_3}$	H_2O above 110°C., per cent
25– 50	3	1.68	1.13	14.3
50–100	3	0.94	0.62	16.4
100–150	3	0.37	0.22	19.0

Aggregates in Soils.—Baver (3) writes:

Soil structure is usually defined as the arrangement of the soil particles. This concept, however, requires a clear understanding of the word "particles." As far as structure is concerned, soil particles refer not only to the individual mechanical elements (primary particles), such as sand, silt, and clay, but also to the aggregates or structural elements (secondary particles) that have been formed by the aggregation of smaller mechanical fractions.

The amount of aggregates, or secondary particles, can readily be determined by an aggregate analysis that rests on elutriation methods. Aggregate analyses of soil samples collected by Leonard and Jenny along the 11°C. annual isotherm have been made by Baver. He determined first the percentage by weight of the sum of primary and secondary particles larger than silt size (>0.05 mm. diameter). Then the samples were subjected to an ultimate mechanical analysis whereby all aggregates were destroyed, thus permitting the determination of the large (>0.05 mm.) primary particles. The difference between the two results was equivalent to the fraction of large aggregates present. The results shown in Fig. 74 indicate a pronounced relationship between percentage of aggregates and rainfall. The nature of the curve is in harmony with the previously mentioned clay and colloid curves. Under low rainfall, the total clay content of the soil is low, thus precluding the formation of a large number of secondary particles. As rainfall becomes greater, weathering becomes more intense, clay formation increases, and aggregation is favored, particularly in the presence of abundant organic matter.

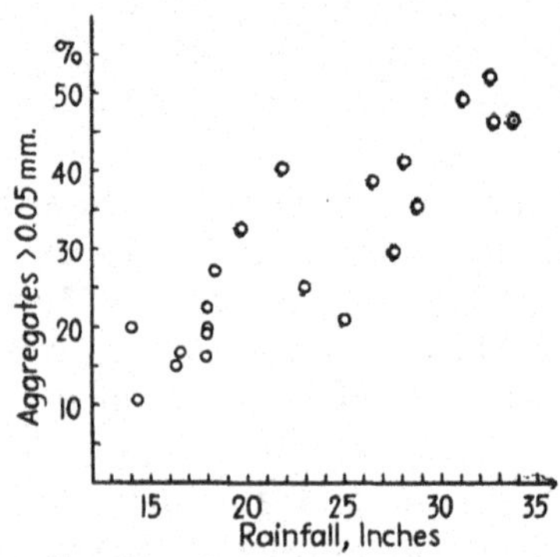

Fig. 74.—Percentage of aggregates larger than 0.05 mm. found in surface soils along the 11°C. isotherm. (*Courtesy of L. D. Baver.*)

Aggregation data are a valuable aid for the elucidation of problems relating to soil structure.

Effectiveness of Rainfall and Intensity of Soil Formation.—The soil property-moisture functions so far evaluated cover mainly the climatic transition interval and are of a linear, logarithmic, or exponential form. All data indicate that over a wider moisture range the curves are of the sigmoid type pictured in Fig. 75. Mathematically speaking, the *effectiveness of* 1 *in. of mean annual rainfall* is highest at the point where the value of the first differential coefficient $f'(m)$, reaches a maximum. For most properties, this particular point appears to occur in the semiarid zone (inflection point in Fig. 75). In regard to the effect of increasing rainfall on acidity (pH), however, the critical point lies in the semihumid region (Fig. 71). The sigmoid curve further indicates that the more abundant the rainfall the smaller its efficiency per unit of moisture.

The influence of rainfall upon soil development is not equally pronounced for all constituents involved in the process. This contention is well illustrated in Fig. 76, which shows the relative change of certain soil properties with rainfall. These curves were obtained by arbitrarily setting the value of the soil property at a rainfall of 15 in. equal to one. Two groups of curves can be distinguished clearly:

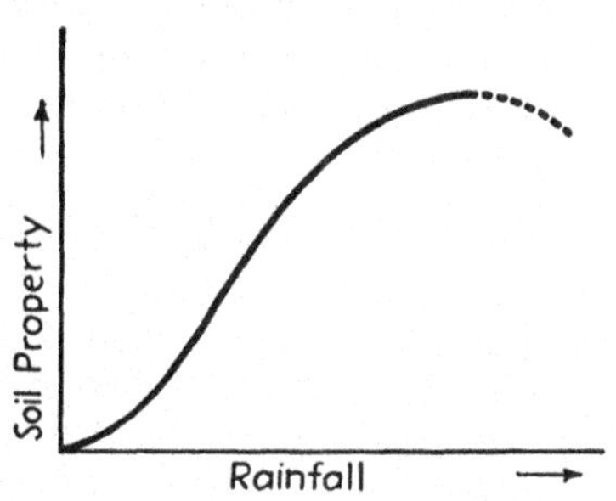

Fig. 75.—*Sigmoid curve* illustrating the general shape of a number of soil property-rainfall functions extending from arid to humid regions.

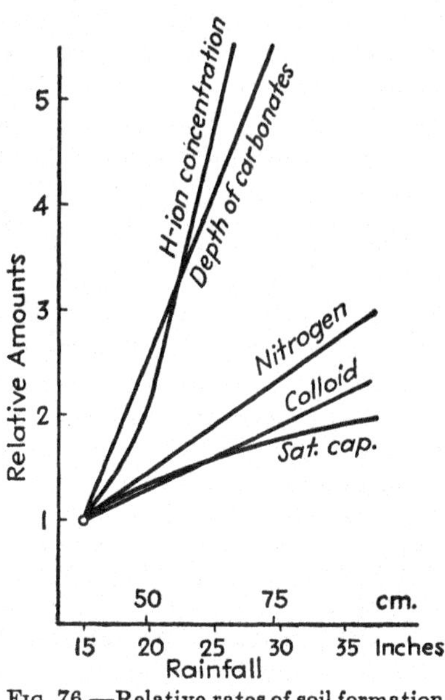

Fig. 76.—Relative rates of soil formation in the arid-humid transition region.

High rate of variation. Hydrogen ion concentration, depth of carbonate horizon.

Low rate of variation. Nitrogen and organic matter, colloidal clay, saturation capacity.

The hydrogen ion concentration and the depth of the carbonate horizon react most sensitively to rainfall; next come nitrogen and organic matter and finally clay and saturation capacity.

B. TEMPERATURE AS A SOIL-FORMING FACTOR

1. Discussion of Temperature Criteria

With the exception of the mountainous districts, the mean *annual air temperatures* are highest in equatorial regions and gradually decline toward the poles. For example, the mean annual temperature at Batavia in the Dutch East Indies is 78.8°F., whereas at Verkhoyansk in Siberia it is 2.7°F. Key West, Fla., reports 76.8°F. and Devils Lake, N. D., 37.2°F.

The mean annual temperature is an abstraction and for many purposes an inadequate index of the annual heat conditions of a given locality. Nevertheless, in soil investigations carried out in many parts of the world, mean annual temperatures have proven quite satisfactory. Significant quantitative correlations between soil properties and mean annual temperatures were obtained, provided regions with similar seasonal temperature trends were selected.

Regions with little or moderate seasonal variations of temperature and rainfall are said to have *oceanic* climates. *Continental* climates are characterized by very hot summers and extremely cold winters. In refining soil property-climate correlations, the annual march of temperatures may prove of importance and could then be treated as a separate soil-forming factor.

Thornthwaite (77) has developed the concept of the temperature efficiency index (*TE* index), which is essentially the accumulated sum of monthly temperatures. In analogy with the definition of precipitation effectiveness (see page 111) the *TE* index (I') is defined as follows:

$$I' = \sum_{n=1}^{n=12} \left(\frac{T - 32}{4} \right)_n \tag{28}$$

I' = annual temperature efficiency index,
T = average monthly temperature in degrees Fahrenheit. The value 32 is used for temperatures below 32°F.,
n = number of months.

The poleward limit of the tundra has a *TE* index of 0, whereas the poleward limit of the tropical rain forest and savanna assume a value of 128.

Soil and Air Temperatures.—As in the case of rainfall measurements, the official temperature readings are taken several feet above the ground; consequently actual soil temperatures may differ profoundly from the data published by the meteorological stations. However, air temperatures and soil temperature are as a rule functionally interrelated (70). Detailed accounts relating to soil temperature are to be found in modern textbooks on soil physics. A summary of information on soil temperatures in the United States has been published in the *Monthly Weather Review* (Vol. 59, pages 6 to 16, 1931).

Van't Hoff's Temperature Rule.—Temperature as a soil-forming factor has long been neglected, but today its importance in the general scheme of soil genesis can no longer be denied. The significance of temperature is easily appreciated by taking into consideration van't Hoff's (83) temperature rule, which can be formulated as follows: *For every 10°C. rise in temperature the velocity of a chemical reaction increases by a factor of two to three.* The rule holds for a large number of chemical reactions, particularly slow ones, and applies equally well to numerous biological phenomena. The term "rule" suggests that exceptions occur. These indeed are plentiful. The rate factor for the 10°C. interval is not constant, and coefficients below 2 and above 3 are quite common. This is due to the fact that the velocity-temperature function often is of a complicated exponential nature and, in the case of living systems, usually exhibits a maximum. Van't Hoff's rule itself is of an empirical nature, but the general observation that chemical reaction rates increase exponentially with rising temperature is supported by theoretical considerations.

Ramann's Weathering Factor.—Ramann (60) expressed the opinion that chemical weathering consists essentially of a hydrolytic decomposition of the silicates. On the basis of this concept, the degree of the dissociation of water becomes of paramount importance. Ramann emphasizes the connection between dissociation of water and temperature and quotes the data listed in Table 33.

TABLE 33.—DISSOCIATION OF WATER INTO H AND OH IONS (*Ramann*)

Temperature, °C.	0	10	18	34	50
Temperature, °F.	32.0	50.0	64.4	93.2	122.0
Relative degree of dissociation of water	1	1.7	2.4	4.5	8.0

At soil temperatures below 0°C., chemical reactions in the soil practically stop; therefore, only temperatures above freezing should be dealt with in soil-formation studies. Not only the absolute soil temperature but also the length of the annual weathering period have to be considered. Ramann arrives at a weathering factor by multiplying the annual number of days having temperatures above freezing by the relative degree of dissociation of water (Table 34). In tropical regions, weathering proceeds three times faster than in temperate zones and nine times more rapidly than in the Arctic.

TABLE 34.—RAMANN'S WEATHERING FACTOR

Region	Average soil temperature	Relative dissociation of water	Number of days of weathering	Weathering factor	
				Absolute	Relative
Arctic.......	10°C.	1.7	100	170	1
Temperate..	18°C.	2.4	200	480	2.8
Tropical....	34°C.	4.5	360	1,620	9.5

Although the data in Table 34 are open to criticism, the underlying idea of combining in some way the effect of temperature, preferably in the form of van't Hoff's law, and the length of the "weathering season" undoubtedly is a fruitful one.

2. RELATIONSHIPS BETWEEN SOIL PROPERTIES AND TEMPERATURE

Depth of Weathering.—A familiar observation of the early soil scientists was that in humid warm regions the rocks had weathered to much greater depths than in the cold zones. In northern Europe, in the Alps, in the Northern United States, and in Canada, the thickness of the soil is usually expressed in inches or centimeters. Rarely does the *solum* exceed a few feet in depth. In contrast, the weathered mantle of subtropical and tropical regions achieves huge thicknesses, and often one must dig for many feet or yards before the fresh rock is exposed. Depths of from 130 to 160 ft. have been frequently observed, and Vageler reports a case of 1,312 ft.

Soil Color.—In humid regions of the cold and temperate zones, the soils are predominantly of grayish color, which is often modified toward black or brown, according to the amount and nature of organic matter and iron hydroxide. In young soils, the color of the parent material is strongly reflected in the color of the soil.

Many tropical soils, especially those derived from igneous and metamorphic rocks, are characterized by brilliant yellow and dark-red colors. However, localized areas of brown, gray, and even black colors also have been described. On certain limestones, the reddish-colored soils extend northward into the temperate region, *e.g.*, the famous Sicilian red earth or the limestone soils of the Ozark plateau.

If the color of the soil appears to be associated with a given climate rather than with the parent rock or specific local conditions, it is customarily spoken of as a climatic soil color. In soil literature, such terms as brown forest soils, gray soils, yellow soils, red soils always imply a climatic origin of color; in fact, the names correspond to the climatic soil types of humid regions. The same holds true for the chestnut soils and the black earths of the semiarid areas.

Attempts have been made to link the color of geological deposits with the nature of the climate at the time of the formation of the sediments. In particular, red strata are commonly assumed to be the result of tropical weathering. This assumption certainly is not justified in all cases, and the conclusion should be substantiated by further investigations.

Leaching of Bases.—For 62 soils derived from igneous and metamorphic rocks, the leaching value β was calculated by dividing the molecular ratio $\frac{K_2O + Na_2O}{Al_2O_3}$ of the most leached horizon by that of the parent rock (Table 35). The average β values decrease consistently from the temperate to the tropical soil groups, indicating an increase in the stage of weathering from north to south. The difference between the β values of the podsolized soils on the one hand and the laterites on the other amounts to 83 per cent. It is, on statistical grounds, beyond question.

The divalent cations behave in like manner. If the molecular ratio $\frac{CaO + MgO}{Al_2O_3}$ of the most leached horizon is divided by that of the parent material, the leaching value β_{II} is obtained. As may be seen from the last column in Table 35, the β_{II} values reveal significant leaching tendencies in the three major heat zones. An increase in chemical decomposition is indicated by a decrease in the value of β_{II}.

It would be erroneous to attribute the entire change in the leaching values solely to alterations in temperature conditions.

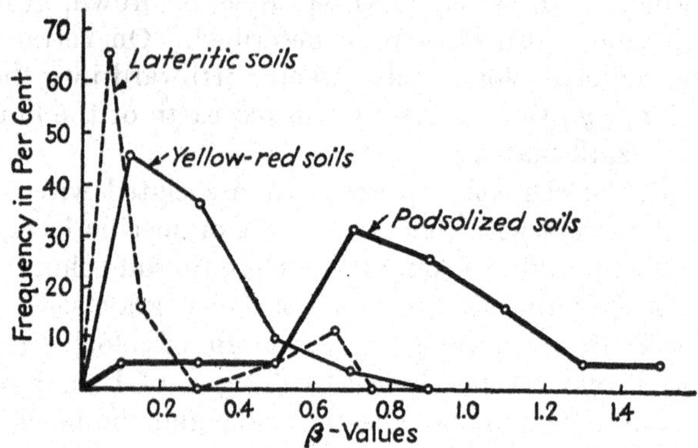

FIG. 77.—Distribution of leaching values β of podsolic soils, yellow-red soils, and lateritic soils.

In the tropics, high precipitation and long durations of weathering have also shared in the development of highly leached soils.

TABLE 35.—LEACHING VALUES FOR MONO- AND DIVALENT IONS AS RELATED TO TEMPERATURE BELTS
(Fusion analyses of soils derived from igneous and metamorphic rocks)

Temperature belt	Climatic soil types	Leaching values	
		(K + Na) β	(Ca + Mg) β_{II}
Cold and temperate..	Podsolized soils	0.822 ± 0.073	0.725 ± 0.082
Subtropical..........	Yellow-red soils	0.278 ± 0.035	0.249 ± 0.062
Tropical.............	Laterites	0.141 ± 0.050	0.034 ± 0.009

Functional Relationships between Temperature and Soil Properties

Functions between soil properties and temperature are expressed by the following form of the general equation for soil-climate relationships:

$$s = f(T)_{m,o,r,p,t,\ldots} \tag{17}$$

T represents temperature as the independent variable. Moisture (m), organisms (o), topography (r), parent material (p), and time (t) are to be kept constant.

The United States again offers an ideal territory for soil formation-temperature investigations, since broad soil belts with comparable moisture conditions run from north to south through areas of varied temperatures. In certain areas, parent material, vegetation, and topography are sufficiently uniform to enable an approximate solution of the problem.

A difficulty requiring some thought is the selection of suitable moisture criteria. Certain investigators are content with annual precipitation values, disregarding the fact that a given amount of rainfall is not so effective in the tropics as it is in the north temperate region, owing to a great difference in evaporation. The slopes of soil property-temperature functions will vary greatly according to the nature of the moisture index selected. In the studies herein reported, the precipitation-evaporation ratio, or, more specifically, its substitute, the NS quotient, has been chosen as a basis for the selection of comparable moisture regions.

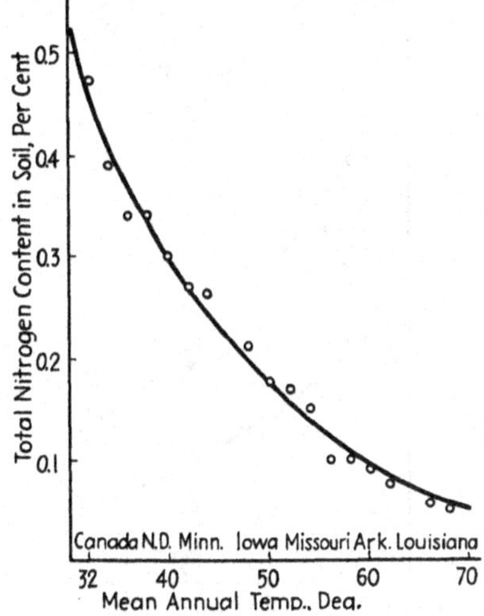

FIG. 78.—Showing average total nitrogen content of the soil as related to the mean annual temperature (Fahrenheit) in the semihumid region.

Nitrogen and Organic Matter as a Function of Temperature.—The area east of the Rocky Mountains has been surveyed to secure quantitative information on the effect of temperature on the amount of nitrogen and organic matter in the soil. As virgin soils approach maturity, an equilibrium is attained between the accumulation of organic matter by vegetation and its destruction by microorganisms. According to the law of van't Hoff, it is to be expected that temperature exerts a decided influence on the balance of the production and decomposition of organic matter. Laboratory experiments by Wollny (90) and more recently by Waksman and Gerretsen (86) support this idea. Of the numerous nitrogen-temperature functions found in the literature, two typical examples are reproduced in Figs. 78 and 79.

The data for the semihumid region (*NS* quotient range 280 to 380) are presented in Fig. 78. Each point represents the average nitrogen content of the surface layer (from 0 to 7 in.) collected from extensive areas of upland soil types. Individual soil nitrogen analyses naturally would exhibit greater fluctuations than the means. The soil samples were taken, for the most part, from cultivated fields; however, in spite of the fact that cultivation disturbs the natural organic-matter equilibrium, a pronounced negative correlation between nitrogen and temperature is brought to light. Gillam (22) observed that in the Great Plains area the relative pigment content and the relative humus content also decrease with increasing temperature.

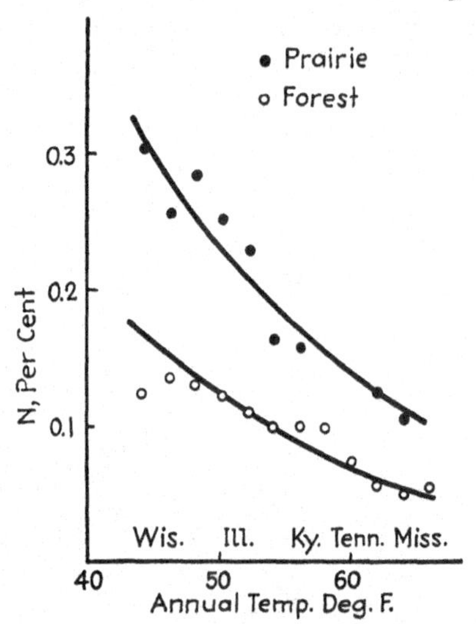

FIG. 79.—Nitrogen-temperature relation in humid grassland (upper curve) and humid timber soils for silt loams.

The data of Fig. 79 for the *humid region* (*NS* quotient 300 to 400 for the area selected) illustrate a similarly pronounced effect of temperature on soil nitrogen. The principal upland soils of the silt loam type were separated according to the original vegetational cover. For both the prairie and the timber soils, the total nitrogen content decreases rapidly from north to south, although the rates of decline are different. Similar curves were obtained for the organic-matter content of originally timbered soil of the Eastern United States. In regard to virgin forest soils, Fisher (24) writes the following:

If one considers the characteristic forest soils from the subarctic regions of eastern Canada to the Appalachian regions of Kentucky and Tennessee, it will be noticed that the most conspicuous differences lie in the depth and condition of the sum total of organic material that remains above the mineral soil. In the northern extremes, these layers are excessively deep and slow to decompose. In the southern example, they are shallow and subject to rapid decomposition.

The *distinguishing features* of the nitrogen and organic-matter functions may be summed up in the following way:

a. Within belts of uniform moisture conditions and comparable vegetation the average nitrogen and organic-matter contents of the soil decrease as the annual temperature rises. The relationship is exponential and may be described satisfactorily by equations of the type

$$N = Ce^{-kT} \tag{29}$$

where N represents the total nitrogen or organic-matter content of the surface soil, T the temperature, and C and k are constants.

b. The following empirical rule applies: for each fall of 10°C. in annual temperature, the average total nitrogen and organic-matter content of the surface soil increases from two to three times, provided that the annual precipitation-evaporation ratio is kept constant. In warm climates, the decomposition of vegetable matter is accelerated; in cool regions, accumulation is favored.

The Organic-matter Problem in the Tropics.—The nitrogen-temperature investigations tend to clarify some of the inconsistencies and contradictions of discussions of soil humus of tropical regions. Certain authors maintain that tropical soils are necessarily low in organic matter because of the high temperatures that hasten the decomposition of plant residues (11). On the other hand, some writers seem to be fully convinced that soils in low latitudes are exceptionally rich in humus as a consequence of the tremendous production of organic substances by the luxurious flora. Published nitrogen analyses of tropical soils lend weight to both arguments.

Reasoning based on van't Hoff's law should deal with soil types of comparable moisture and kinds of vegetation. In the humid United States the annual yield of vegetable matter does not vary excessively, being of the order of from 1 to 3 tons per acre. It is probably for this reason that the temperature influence manifests itself in such an outspoken manner and may be closely represented by negative exponential functions. Extending the curves in Figs. 78 and 79 to still higher temperatures, it appears logical to conclude that, for comparable conditions in the tropics, nitrogen and organic matter, indeed, should be at a very low level.

If, however, the type of vegetation differs materially from that of North America and produces yields of from 40 to 80 tons per acre, as is conservatively estimated by Vageler, the situation takes on an entirely different aspect. A moderate accumulation of nitrogen is not unlikely, but, as Vageler (82) points out, under similar conditions, soils of temperate and cold regions would acquire organic-matter layers of several yards in thickness.

Mohr (55) has attempted to illustrate diagrammatically the position of the humus equilibrium assuming that temperature and production and destruction of organic matter vary simultaneously. Mohr's conclusions read as follows: "In well-aerated soils of the humid warm tropics with average temperatures over 77°F. (25°C.) humus cannot maintain itself nor can it accumulate."

In contrast to the tropics, it is of interest to note that in the high altitudes of the Alps, where the annual temperature is below 32°F. (0°C.), the microbiological activity is delayed, and the surface soils contain from 20 to 30 per cent organic matter, in spite of sparse vegetation.

Clay-temperature Functions.—In the Eastern part of the United States an extensive region of igneous and metamorphic rocks extends from Maine to Alabama. The mean annual temperature of this belt ranges from 6 to 19°C. For the past 40 years, extensive studies of this area by soil scientists of the U. S. Department of Agriculture have resulted in the publication of numerous maps and soil reports. The latter contain the results of over 1,000 mechanical analyses of a great variety of soils. Because of strictly standardized methods of sampling and analytical technique, these data are admirably suited for a quantitative study of the effect of temperature on the clay content of the soil. Unlike sedimentary deposits, the igneous rocks do not contain clay particles; hence the clay content of residual soils derived from parent materials of magmatic origin is the result of weathering and thus furnishes an index of the intensity of rock decay.

Of all the data available, about 150 analyses of surface and subsoils are comparable; that is to say, they reasonably satisfy the requirements of constancy of all soil-forming factors except temperature. By combining the surface and subsoil analyses in a suitable manner, the *average clay content* (particles smaller than 5μ) *of the soil profile* to a depth of 40 in. has been calculated.

It can easily be demonstrated that the northern soils contain considerably less clay than the southern soils, the latter having a clay level about 2.3 times higher than that of the former

TABLE 36.—CLAY LEVELS OF NORTHERN AND SOUTHERN LATITUDES
(Depth 0 to 3 ft.)

Latitude, degrees	Temperature range, degrees centigrade	Geographic region (states)	Number of profiles	Average clay content, per cent
35–45	6–12	New England states, New York, Pennsylvania, New Jersey, Maryland, North Carolina (glaciated and unglaciated region)	39	16.3 ± 1.18*
30–35	16–19	South Carolina, Georgia, Alabama	33	38.0 ± 1.30

* Mean error.

TABLE 37.—CLAY LEVELS FOR VARIOUS LATITUDES AND TEMPERATURE INTERVALS
(Glaciated areas excluded)

Latitude, degrees	Temperature range, degrees centigrade	Geographic region (states)	Number of profiles	Average clay content, per cent
35–41	10–13	Pennsylvania, New Jersey, Maryland, Virginia, North Carolina	38	25.6 ± 1.24*
30–35	16–19	South Carolina, Georgia, Alabama	33	38.0 ± 1.30

* Mean error.

(Table 36). In this comparison, the soils of the glaciated part of the Eastern United States have been included, a procedure that may be open to objection. Inasmuch as the glacial soils have had less time for development than the soils of the southern Piedmont Plateau, it is possible that the clay differences are not only the result of climate but also a consequence of the youthfulness of the northern soils.

In Table 37, only soils of unglaciated areas have been grouped. It may be clearly seen that the clay differences still persist. The unglaciated soils of latitudes from 35 to 41° contain but 26 per cent clay, whereas farther south in the latitude range from 30 to 35° the average clay level is 38 per cent. This remarkable rise may be taken as direct evidence of climatic influences.

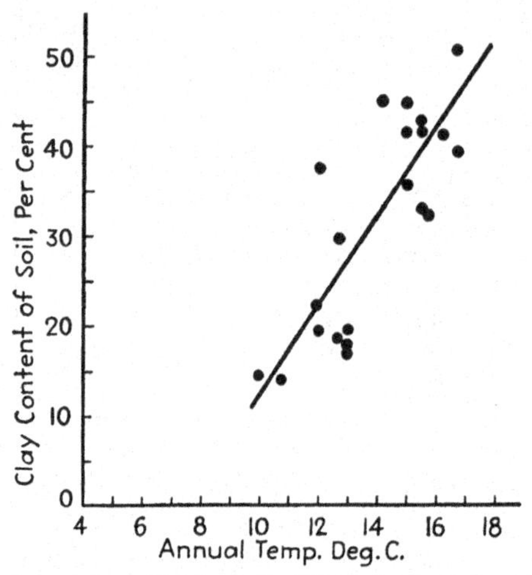

FIG. 80.—The clay content of soils derived from basic rocks varies in an orderly manner from north (low temperatures) to south (high temperatures). Every dot denotes the average clay content of a single soil profile to a depth of 40 in.

If clay analyses are classified according to the type of unweathered rock (granite, gneiss, basic rocks, etc.) and plots made of clay content against temperature for each group of rocks separately, curves that are straight-line functions are obtained. An example is given in Fig. 80, which shows the correlation for 21 soils derived from basic rocks, mainly diorite and gabbro. The function may be expressed as

$$\Gamma = 4.94T - 37.4 \tag{30}$$

Γ denotes the average clay content in per cent and T the annual temperature in degrees centigrade. The relationship is statistically significant, the correlation coefficient having a value of +0.814. The individual data have been arranged in Table 38.

A careful study of Table 38 shows that for the soil samples collected the requirement of constant moisture is not strictly fulfilled, because neither rainfall nor NS quotient is equal for all localities. There is a general downward trend in the effective humidity (*NS* quotient) with increasing temperature. In order to obtain a more accurate picture of the role of temperature, some

TABLE 38.—DATA FOR THE PARENT MATERIAL GROUP BASIC ROCKS
(Climatic data represent annual values)

Number	State	County or area	Temperature, degrees centigrade	Rainfall, inches	*NS* quotient	Soil type	Clay content, per cent
1	New Jersey	Bernardsville	10.0	48.4	400	Montalto silt loam	14.4
2	New Jersey	Belvidere	10.8	47.9	400	Montalto silt loam	13.9
3	New Jersey	Trenton	12.0	48.8	437	Cecil loam	22.3
4	Maryland	Montgomery	12.1	38.4	350	Conowingo silt loam	19.5
5	Virginia	Leesburg	12.2	40.3	450	Iredell clay loam	37.6
6	Maryland	Howard	12.8	42.7	316	Montalto clay loam	29.7
7	Maryland	Howard	12.8	42.7	316	Mecklenburg loam	18.7
8	Maryland	Baltimore	13.0	42.7	316	Mecklenburg loam	17.2
9	Maryland	Baltimore	13.0	42.7	316	Iredell silt loam	19.3
10	Maryland	Baltimore	13.0	42.7	316	Conowingo silt loam	17.7
11	North Carolina	Caswell	14.3	46.0	400	Iredell sandy loam	45.3
12	North Carolina	Cabarrus	15.1	46.9	300	Iredell fine sandy loam	41.6
13	North Carolina	Cabarrus	15.1	46.9	300	Mecklenburg clay loam	45.0
14	North Carolina	Cabarrus	15.1	46.9	300	Mecklenburg sandy loam	35.8
15	North Carolina	Mecklenburg	15.6	46.9	300	Iredell fine sandy loam	32.9
16	North Carolina	Mecklenburg	15.6	46.9	300	Mecklenburg clay loam	41.9
17	North Carolina	Mecklenburg	15.6	46.9	300	Mecklenburg loam	42.7
18	North Carolina	Gaston	15.7	46.9	300	Iredell clay loam	32.6
19	South Carolina	Abbeville	16.3	47.6	300	Iredell clay loam	41.5
20	South Carolina	York	16.8	49.8	300	Iredell clay loam	39.7
21	Georgia	Meriwether	16.8	48.9	300	Davidson clay	51.1

sort of moisture correction becomes necessary. This is accomplished by arbitrarily adjusting all clay values to an *NS* quotient value of 400 on the basis of the clay-moisture relations shown in Figs. 67 and 68. The corrected curves are exponential in character. The curves fitted according to the method of least squares follow the equation

$$\Gamma = Ce^{kT} \tag{31}$$

The letters C and k represent constants. Seven such correlations for different groups of parent material have been established.

Two of the correlation curves are graphically presented in Figs. 81 and 82.

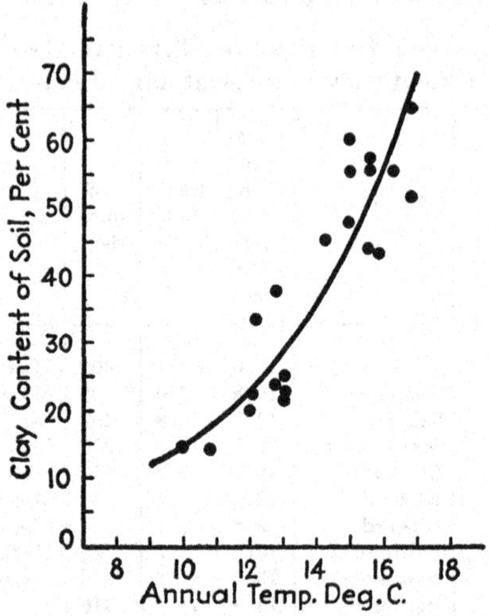

FIG. 81.—Clay-temperature function for soils derived from basic rocks. Data adjusted to constant moisture expressed as *NS* quotient 400.

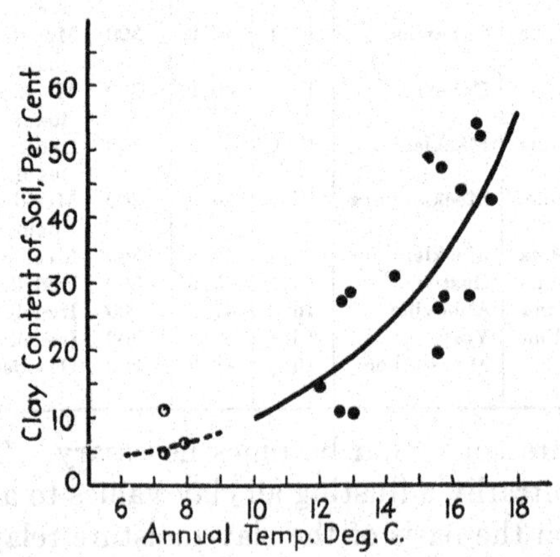

FIG. 82.—Clay-temperature function for soils derived from granites; constant moisture *NS* quotient 400. The black-white circles represent soils from morainic materials and have not been included in the calculation of the curve.

Generally speaking, the following outstanding features are common to all curves.

a. The clay content of comparable soils is greater the higher the annual temperature.

b. Under constant moisture conditions (*NS* quotient = 400), the clay-temperature function is of an exponential form.

The exponential form of the above relationships is in approximate agreement with van't Hoff's temperature rule.

It should be borne in mind that the slopes of the curves depend on the specific moisture criteria employed. Different methods of approach may lead to different constants, but these would not alter the established principle of the temperature influence on clay formation.

In this connection, the clay analyses of soils of arctic regions collected by Dutilly in northern Canada (17) merit consideration. The clay content (particles smaller than 2μ) was found to be generally small, only three of 22 samples having as much as 20 per cent clay and a majority of the remainder having less than 5 per cent.

Chemical Composition of Clay in Relation to Temperature.—A number of years ago, Robinson and Holmes (65) published a series of clay analyses of American soil types that revealed a slight correlation between composition of soil colloids and rainfall but no correlation between composition of the colloids and temperature. However, when the analyses from similar moisture regions are compared, a temperature effect is brought to light. For areas of constant *NS* quotient (from 300 to 400), the silica-alumina and the base-alumina ratios of colloids extracted from surface soils become narrower as the temperature increases.

For the podsols of the North, the quotient $SiO_2:Al_2O_3$ is from 3 to 4 or even higher, whereas for the red soils of the Southeastern United States, it is less than 2. The $SiO_2:Al_2O_3$ ratio of 2 appears to occur at an annual temperature of 16°C. (60.8°F.), a result that has been confirmed by Baver and Scarseth (4) and also by Davis (15).

The above interpretation of the results of Robinson and Holmes has been attacked by Crowther (14). Supported by more rigorous statistical methods, he denies that temperature is a causal element in the aforementioned trend of the silica-alumina ratio. He maintains that for constant rainfall the ratio becomes wider with increasing temperature. Superficially it would seem that the conclusions of Crowther and those of the author are opposed. It should be noted, however, that the discrepancy is not necessarily a fundamental one. Crowther correlates clay composition and temperature on the basis of constant precipita tion, whereas the above interpretation is made on the basis of

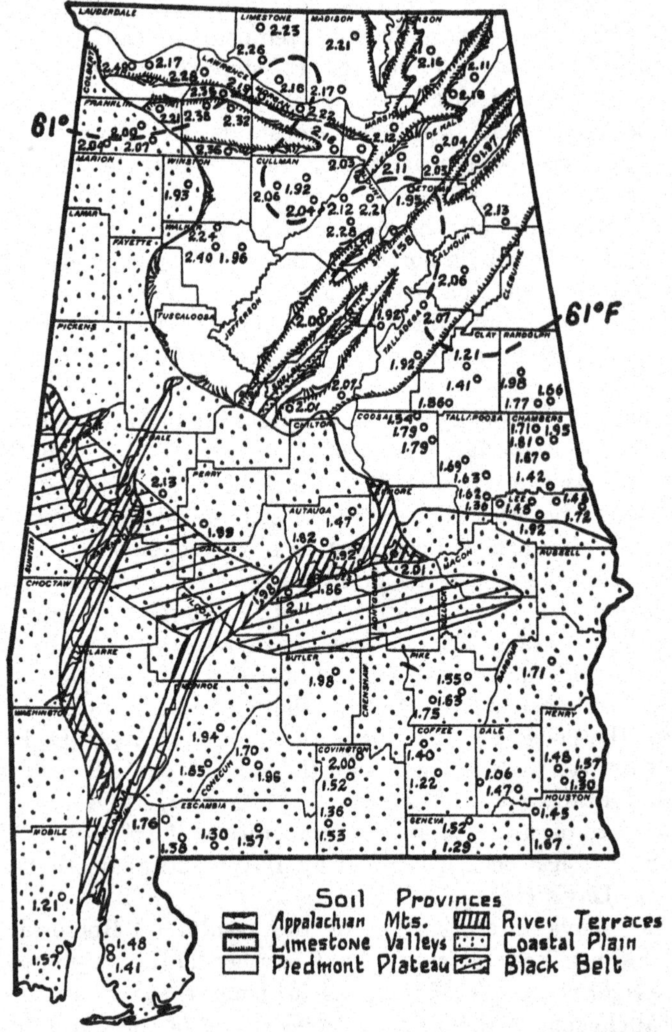

FIG. 83.—Silica-alumina ratios of the colloidal fraction of surface soils from the state of Alabama. (*Davis.*)

constant *NS* quotient, which takes into consideration the rainfall effectiveness as influenced by transpiration and evaporation. Numerous scientists have participated in this controversy, which, at present, is still open (7). The issue is of considerable importance from the standpoint of soil formation because of its close relation to the problem of the causes of lateritic and podsolic weathering.

Figure 83 shows the data obtained by Davis for subtropical soils of Alabama. The map quite satisfactorily depicts the interaction of parent material and climate on the silica-alumina ratio of soil colloids from surface horizons. Irrespective of the type of parent rock, all *sa* values are uniformly low.

TABLE 39.—SiO_2-Al_2O_3 RATIO OF SOIL COLLOIDS FROM ARID REGIONS. SURFACE HORIZONS [*Brown and Byers* (6)]
(Variable parent material)

General location	Annual temperature, °F.	Annual rainfall, inches	$\frac{SiO_2}{Al_2O_3}$
North Dakota	39.8	13.7	4.14
South Dakota	46.0	14.5	5.28
South Dakota	46.2	17.7	3.35
South Dakota	47.2	14.5	3.76
Nebraska	49.9	16.9	4.64
Kansas	54.3	22.8	4.25
Texas	55.8	20.8	3.89
New Mexico	61.2	8.5	4.02
New Mexico	62.7	14.1	4.02
Arizona	68.8	10.5	3.55

In arid regions, the influence of temperature on colloid composition is small, as may be seen from the data of Brown and Byers in Table 39. In spite of enormous variations in temperature, the silica-alumina ratios of the clay particles are wide and fluctuate between 3.35 and 5.28. Brown and Byers attribute the variability of the ratios to variations in the parent material, which, except in the case of a granitic soil from Arizona, consists of sedimentary rock.

Podsol Profiles and Lateritic Profiles.—Proceeding from the Arctic to the equator along lines of high rainfall, the following major climatic soil groups (zonal soils) are encountered:

Skeletal soils,
Tundra soils,
Podsols,
Gray-brown-podsolic soils,

Yellow-red soils,

Lateritic soils and laterites.

These groups are characterized by specific profile patterns, as indicated by both field and laboratory observations. The depth functions of the podsols and the lateritic soils are particularly impressive.

TABLE 40.—CHEMICAL COMPOSITION OF A PODSOL SOIL (BECKET LOAM)
[Collected by W. J. Latimer, analyzed by G. H. Hough and G. E. Edgington (49)]

Horizons	A_1	A_2	B_1	B_2	C
Depth, inches	0–6	6–11	11–13	13–24	24–36
SiO_2, per cent	52.95	83.32	69.60	72.67	77.86
Al_2O_3, per cent	7.04	6.73	9.61	10.32	10.00
Fe_2O_3, per cent	1.08	1.69	3.99	3.58	3.15
MnO, per cent	0.01	0.01	0.01	0.02	0.03
CaO, per cent	0.90	0.54	0.65	0.62	0.54
MgO, per cent	0.15	0.18	0.33	0.41	0.48
K_2O, per cent	2.06	2.89	3.41	3.45	3.79
Na_2O, per cent	0.40	0.46	0.46	0.67	0.55
TiO_2, per cent	0.66	0.90	0.79	0.70	0.53
P_2O_5, per cent	0.13	0.04	0.08	0.08	0.08
SO_3, per cent	0.36	0.13	0.20	0.14	0.13
Ignition loss, per cent	34.40	2.75	11.25	7.27	2.54
Total, per cent	100.14	99.64	100.38	99.93	99.68
N, per cent	1.04	0.05	0.14	0.09	0.02
Clay ($<5\mu$), per cent	3.8	7.0	10.6	9.5	8.9
pH	3.8	3.7	3.9	4.1	4.5
sa	12.8	21.0	12.3	11.9	13.2
ba	0.64	0.72	0.59	0.58	0.60

Podsols.—In vast areas of northern Europe, Asia, and America are found soil profiles that have a conspicuous arrangement of dark, white, and brown horizons. These soils are known as "podsols," a Russian name that means "ashy soil." The following sequence of horizons is generally recognized:

A_0: Magnitude: <1 to 5 in. Consists of forest litter and leafmold. Sometimes subdivided into F* and H† layers.

A_1: 1 to 6 in. thick. Dark-brown humus layer, mixed with mineral soil. Often designated as raw humus. Highly acidic (pH = 4).

* F = fermentation horizon or first decomposition layer.

† H = humified horizon or heavily decomposed layer.

A_2: 1 to 8 in. thick. Grayish-white, bleached horizon. Zone of maximum eluviation.

B: 2 to 15 in. magnitude. Rust-brown horizon (in iron podsols) or chocolate horizon (in humus podsols), depending on the amount of Fe_2O_3 and humus that have accumulated. The B horizon is thicker than the other layers and in most cases is divided into B_1 and B_2. The lower boundary of B_2 is irregular and frequently forms pocketlike intrusions into the substratum. If the sand grains are cemented together by colloidal sesquioxides (Fe_2O_3 plus Al_2O_3) into a hardpan, the zone is called "ortstein."

C: Parent material, mostly light textured.

The chemical analysis of a typical podsol is shown in Table 40. The entire soil is highly acidic, and, in general, chemical alterations within the profile are not very profound. The $\underset{\cdot}{b}\underset{-}{a}$ values, $\frac{K_2O + Na_2O + CaO}{Al_2O_3}$, vary little, whereas the silica-alumina ratio is low in the B horizon and high in the A horizon, a feature that has been emphasized by Harrassowitz. A number of Scandinavian and Russian investigators have shown conclusively that the essential feature of podsol formation consists in the translocation of sesquioxides from A to B.

Illuminating graphic patterns of podsols are obtained by plotting the SiO_2-Al_2O_3 ratios (*sa* values) of the colloid portion as a function of depth. To facilitate comparison, only relative *sa* values are shown in Fig. 84, that is to say, the *sa* value of the colloid from the C horizon is chosen as unity. All *sa* values of the *solum* horizons have been divided by the *sa* value for the C horizon colloid. For convenience, the absolute values of *sa* for the C colloids are indicated on the "unity line."

The pattern is of a strikingly characteristic zigzag nature. All B-horizon values fall below the unity line. All A-horizon values are far above it. The colloid of the A horizons is greatly enriched in SiO_2, whereas that of the B horizon shows a pronounced relative accumulation of Al_2O_3. No colloid-depth functions from strictly basic rock profiles were obtainable, and it remains to be seen whether or not the podsol type pattern of Fig. 84 is generally valid. For the time being, we shall designate the processes that tend to bring about the specific pattern of colloid composition shown in Fig. 84 as podsolization. This definition is purely formalistic and avoids any reference to the still hypothetical mechanism of the podsolization process. The definition merely states that in podsols the silica content of the inorganic colloid fraction is relatively (with respect to Al_2O_3) enriched in the *A* horizons and relatively impoverished in the *B* horizons as compared with the colloid composition of the *C* horizon. The emphasis placed on the nature of the colloid is in line with the present trend in pedology that considers the colloid fraction of the soil a relatively stable yet sensitive product of the interaction of soil-forming factors.

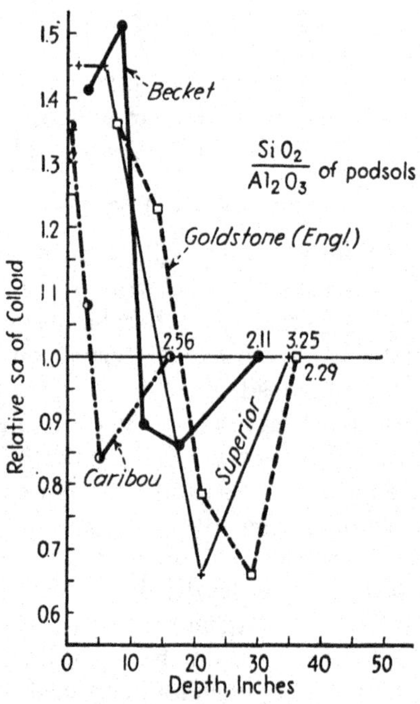

FIG. 84.—Characteristic pattern of podsol profiles. The relative silica-alumina ratio of the inorganic colloidal fraction is plotted as a function of depth.

Laterites and Lateritic Soils.—It is difficult to give a standard description of a laterite profile, in spite of the enormous literature on the subject (Harrassowitz (26), Bennett and Allison (5), Lacroix (45), Fox (19), etc.). The science of pedology originated in northern countries, where soil-formation processes are retarded because of cool winters or dry summers. It is not at all certain that the present-day ideas of American and European pedologists are applicable *in toto* to humid tropical soils, a belief that has been repeatedly expressed by Vageler and more recently by Pendleton

(57). In the tropics we are confronted more than anywhere else with the difficult problem of distinguishing between soil and parent material. If the underlying unweathered rock is considered as the parent material, then the term *solum* must apply to a layer that extends in many cases to hundreds of inches below the surface. If the term "soil" is restricted to the surface portion, *i.e.*, if only the thoroughly weathered portion of rocks is

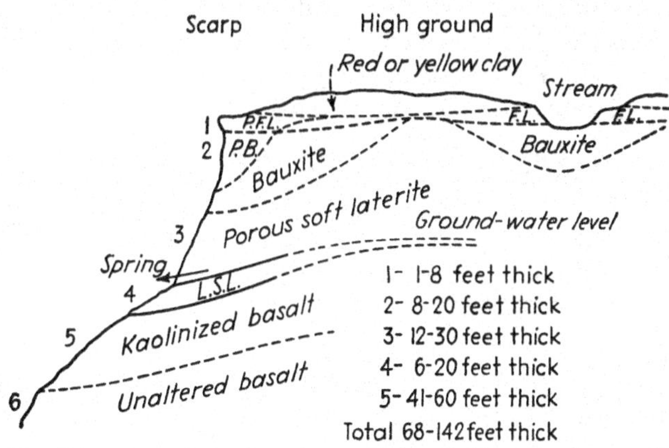

FIG. 85.—Diagrammatic section of part of a typical laterite plateau; *P.F.L.* = pisolitic ferruginous laterite; *F.L.* = ferruginous laterite; *P.B.* = pisolitic bauxite; *L.S.L.* = laminated siliceous lithomarge. (*After Fox, Memoirs Geol. Surv. India* 49, 1923.)

treated as parent material, we come to realize that we have hardly any data on laterite soil profiles, since most investigators extended the *solum* to the bedrock and paid but scant attention to the surface horizons.

Figure 85 is taken from Fox's great memoir on the bauxite and aluminous laterite occurrences in India. It illustrates the sequence of strata from the unaltered basalt rock to the surface layer. As may be seen from the horizon magnitudes listed, the total depth of the "soil" varies from 70 to 140 ft., assuming the parent material to be the unaltered basalt.

In typical laterites, Harrassowitz distinguishes between the following strata:

$\mathfrak{A}$: Humus soil (not always present),
$\mathfrak{B}_2$: Iron crust,
$\mathfrak{B}_1$: Accumulation zone (sesquioxides),
$\mathfrak{Z}$: Decomposition zone (hydrous aluminosilicates),
$\mathfrak{C}$: Fresh rock.

Not all the laterites consist of this complete sequence. Some of the horizons may be absent, or the accumulation of sesquioxides may not constitute a crust. Pendleton urges the restriction of the term "laterite" to the original meaning given by Buchanan in 1830. The term, according to Buchanan (20) and Pendleton, does not refer to a specific soil profile, but merely to a specific horizon or stratum. Typical laterites, as defined by Pendleton, are zones rich in sesquioxides and may be cut into large bricks (= Latin "later") which, upon drying, become hard and resistant. For centuries they have been used for building purposes in many parts of the tropics.

TABLE 41.—DIABASE LATERITE FROM GUINEA (*Lacroix-Boiteau*)

Zones	Surface crust (*zone de concrétion*)	Decomposed rock (*zone de départ*)	Fresh rock (diabase)
SiO_2, per cent	1.30	5.83	51.27
Al_2O_3, per cent	60.19	37.03	12.36
Fe_2O_3, per cent	3.91	31.73	3.29
FeO, per cent	—	—	6.16
MgO, per cent	—	0.06	13.26
CaO, per cent	0.17	0.19	10.66
Na_2O, per cent	—	—	1.60
K_2O, per cent	—	—	0.41
TiO_2, per cent	1.03	1.29	0.70
P_2O_5, per cent	—	—	0.11
H_2O, per cent	32.00	23.02	0.40
Insoluble, per cent	1.40	0.96	—
Total, per cent	100.00	100.11	100.22
$sa = \frac{SiO_2}{Al_2O_3}$	0.367	0.267	7.05
$ba = \frac{K_2O + Na_2O + CaO}{Al_2O_3}$	0.0051	0.0094	1.90

A good example of the severe chemical transformation that rocks undergo in the formation of extreme laterites is provided by the analysis of a diabase laterite from Guinea that was collected by Lacroix and analyzed by Boiteau (Table 41).

Some of the most outstanding chemical characteristics of laterites are revealed in the *ba* values, which indicate that nearly all the bases have been removed. The *sa* values point to a large removal of silica and tremendous relative accumulation of aluminum. The relative accumulation of aluminum is indicative of its high resistance to leaching. These chemical changes that accompany the formation of a true laterite are in excellent accord with the functional relationships observed in Mauritius by Craig and Halais (see page 139).

Like the podsols of the North, the lateritic soils give a specific pattern if the colloid composition is plotted against depth. Although the data are scanty, the few curves shown in Fig. 86 reveal fundamental differences between podsol and lateritic

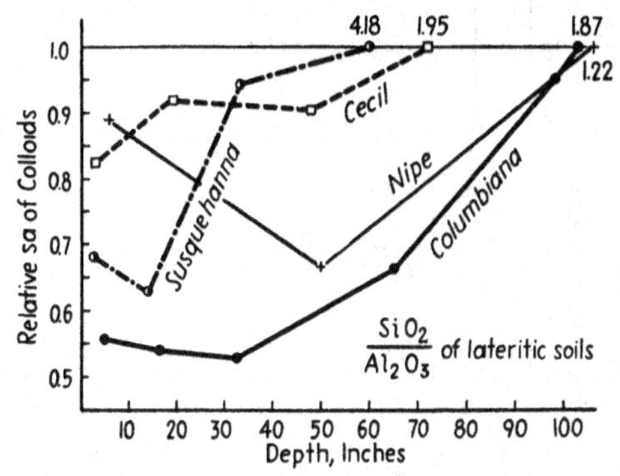

FIG. 86.—Characteristic patterns of lateritic soils. The relative silica-alumina ratio of the inorganic colloid fraction is plotted as a function of depth.

sa depth functions. In striking contrast to the podsol curves of Fig. 84, the curves for lateritic soils lie entirely below the unity line. The reader will note that the colloid curves are restricted to the surface portions of lateritic deposits; in other words, not the fresh rock but its decomposition products have been used as a basis for the calculations. Processes that result in conditions characterized by the curves shown in Fig. 86, shall be designated as *lateritization.* Compared with the *C* horizon the colloid fraction of all horizons of the *solum* is relatively enriched in Al_2O_3.

Leaching versus Podsolization.—The definition of podsolization in terms of *sa* values of colloids is by no means accepted by all pedologists. Particularly the field surveyor uses the term "podsolization" in a much wider sense that is almost synonymous with the terms leaching and eluviation. In his lectures Marbut said: "Podsolization is a leaching process, and the podsol soils are the most thoroughly leached soils of the world insofar as the leaching is done by mere solution." Thorp (79) writes: "Podsolized soils characteristically have a higher percentage of clay in the subsoil than in the surface."

In his Soil Atlas, Marbut chooses the relative accumulation of sesquioxides as a primary criterion of podsolization. For each soil horizon, he calculates the silica-alumina ratio (*sa*) of the entire soil mass instead of the colloid fraction only. Soils with high *sa* values for the *A* horizon and lower values for the *B* horizon, he

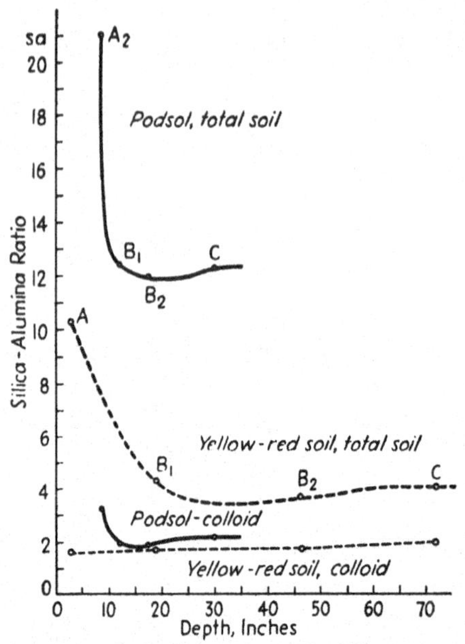

Fig. 87.—Comparison of podsolization criteria based on silica-alumina ratios of total soil mass and of extracted colloid.

considers to be correspondingly podsolized. Although it is quite correct that the classical podsols of the north fit into this routine scheme, the *sa* value of the total mass does not tell the whole truth.

As a case in point, the reader is directed to turn his attention to Fig. 87, which shows the *sa* values of a typical podsol (Beckett loam) and of a typical red soil (Cecil sandy loam). For the soil mass as a whole, the two patterns are remarkably alike. However, if one plots the corresponding *sa* curves for the extracted colloids, a fundamental difference comes to light. The colloid-depth function of the Beckett loam is identical with the typical podsol colloid curves shown in Fig. 84, whereas the Cecil-colloid curve is nearly a straight line with a slight inclination that indicates lateritization. Thus it appears that the *sa* curve of the Cecil soil is mainly the result of mere mechanical migration of unaltered clay particles from *A* to *B*. The high *sa* value of the mass of the *A* horizon is caused by the remaining quartz grains. In the podsol, clay particles also have been translocated, but, in addition, they have been subjected to profound chemical alteration; some of the particles may be entirely new formations of the *B* zone, as assumed by Wiegner (88) and other soil chemists [Gedroiz (21), Mattson (51)].

If the silica-alumina ratio of the entire mass of horizons is selected as a unique podsolization criterion, then many subhumid clay-pan soils might be classified as podsolic soils. A danger exists in the definition of podsolization as being simply leaching or eluviation, since hardly a soil exists that would not, under this too liberal interpretation, be considered as podsolized. The concept would thus lose its significance as a fundamental classification criterion. The extensive podsol studies of Mattson (51) and his associates emphasize the need for supplementing field observations with detailed chemical investigations.

C. COMBINATIONS OF MOISTURE AND TEMPERATURE INFLUENCES

Probably no issue in soil science has been more disputed than the postulate of the dominance of climate in soil formation. The arguments that have been advanced to prove or disprove the contention have been numerous, and even today opinion is divided. Extreme views such as those of Lang on the climatic side and of Van Baren on the geological side are still defended. It must be admitted that undue claims have been made by both the adherents and the opponents of the climatic theory. Fortunately certain investigators, notwithstanding criticism from both sides, have impartially collected quantitative soil data in order to ascertain the true role of climate. The data presented in previous chapters leave no doubt that both rainfall and temperature play a significant part in directing soil-forming processes and in determining certain features of the soil.

A satisfactory evaluation of the influence of climate on soil formation is often complicated by the interplay of moisture and temperature. In certain regions, notably in eastern Europe, and in mountainous districts, the isotherms and moisture isopleths

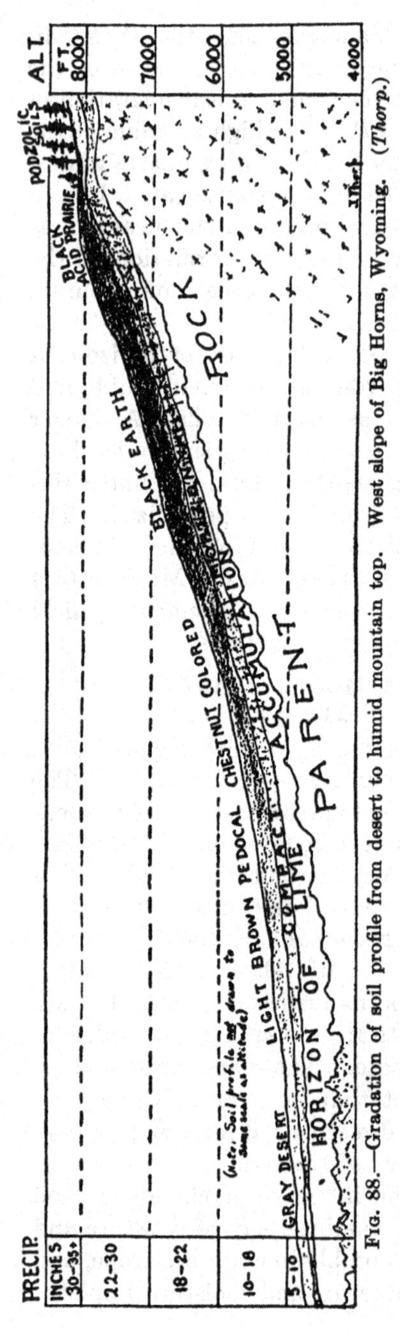

FIG. 88.—Gradation of soil profile from desert to humid mountain top. West slope of Big Horns, Wyoming. (*Thorp.*)

run nearly parallel to each other. It is therefore difficult to ascertain the individual part played by rainfall on one hand and by temperature on the other, although the combined effect may be marked.

Soil Zonality in Russia.—In Table 42 are shown the features of a meridional transect through the European part of U.S.S.R. (18). The soil properties change in a regular manner, and definite soil zones that grade into each other can be observed. The north is characterized by tundra with shallow layers of poorly decomposed plant remains, whereas in the south light-gray soils with occasional salt accumulation (alkali soils) are dominant. Temperature rises regularly from north to south, and rainfall declines from latitude 60° to latitude 45°. This particular association of climatic elements holds good for Russia but not necessarily for other countries. In consequence, the Russian sequence of soil types lacks general applicability.

Mountainous Regions.—According to the laws of climatology, an increase in altitude is accompanied by a falling temperature and a rise in precipitation. Therefore, the sequence of soil zones along a mountain slope should be similar to the south-north transect in Russia. The diagrammatic representation of the soils of the Big Horn Mountains in Wyoming (Fig. 88) by Thorp (78) illustrates this contention in a convincing way. At an altitude of 4,000 ft., gray desert soils and light-brown soils with characteristic lime accumulations are encountered. At higher altitudes, one observes a darkening of the surface soil and a fading out of the lime zone. Clayey subsoils appear. At 8,000 ft., podsolic features with gray and rust-brown horizons may be seen.

TABLE 42.—MERIDIONAL SOIL TRANSECT THROUGH THE EUROPEAN PART OF U.S.S.R. (*Adapted from Filatov and Nekrassov.*)

Selected soil characteristics		Low in humus, neutral or alkaline reaction, salt accumulations	Rich in humus to considerable depth, carbonates, neutral reaction	Dark surface, gray and brown horizons, acid reaction	Poorly decomposed plant remains. Subsoil frozen permanently
Regional soil types		Chestnut soils → ; ← Arid brown soils	Chernozem; Forest steppe; ← Steppe soils →	← Podsolized soils	← Tundra
Climate	July temperature, °C.	25 24 23	22 21 20	19 18 17	<17
	Rainfall, cm. May, June, July	<10	10-15	15-20 20-30 >30	15-20
Locations	Longitude	← about 44° E.	→	← about 40° E.	→
	Latitude	45°	50° 55°	60°	65° N

Sievers and Holtz (74), working in the state of Washington, found that soil nitrogen and organic matter increase with altitude. At 1,000 ft. above sea level, the total nitrogen content was 0.05 per cent. At about 2,000 ft., it was in the neighborhood of 0.1 per cent (surface soils).

Figure 89 depicts the results for virgin soils obtained by Hockensmith and Tucker (30) in the Rocky Mountains in Colorado. These authors state that

The nitrogen content increases with increase in elevation, although above an elevation of 10,000 ft. (3,050 m.) it is noted that the nitrogen content varies considerably. There is great variation in the local climate at these elevations. The high wind velocity causes the snow to drift considerably. Some of these snowdrifts remain the whole year. The samples that were obtained from peaks exposed to the wind and weather were always lower in nitrogen than the soil samples taken at the same elevation, but from the protected areas between the peaks. Vegetation is scarce.

Recently, Hardon (25) has reported excellent soil organic matter and nitrogen relationships in the virgin tropical forests of the mountains of southern Sumatra. Unlike Hockensmith, he

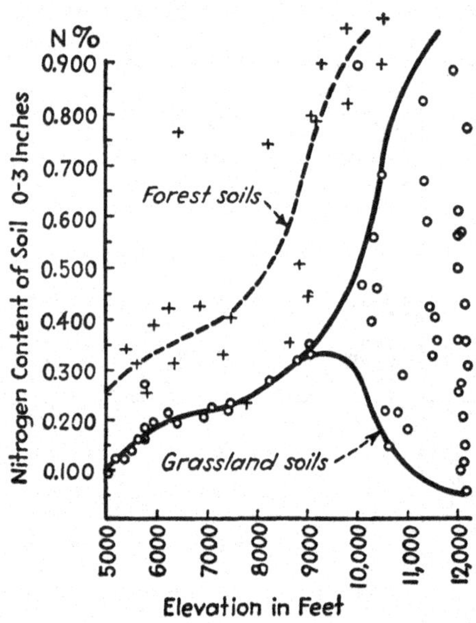

FIG. 89.—Increase of total soil nitrogen with altitude in the Rocky Mountains (*Hockensmith and Tucker.*) Above 9,000 ft. (2,750 m.) the grassland curve resolves into a fanlike scattering owing to erratic growth of vegetation.

has not included in his studies the A_0 horizon, which consists of half-decomposed leaves. With increasing altitude, both organic matter and nitrogen content rise in exponential fashion (Fig. 90). With the accumulation of humus, the carbon-nitrogen ratio becomes wider. The carbon-nitrogen curve exhibits a minimum at about 1,000 ft. above sea level (Fig. 91).

In the mountains of Hawaii, nitrogen and carbon contents of the soil also are positively correlated with altitude, according to Dean (16). In the forests of the Carpathian Mountains in Czechoslovakia, the acidity of the surface layer (from 0 to 8 in.) of podsols increases with altitude. At 1,300 ft., the pH of the

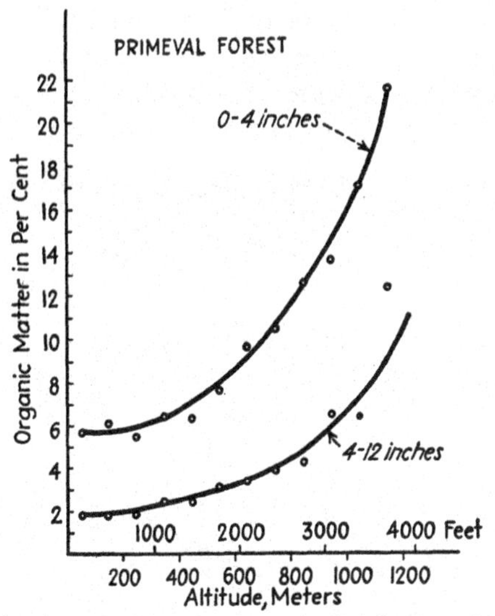

FIG. 90.—Organic matter content of tropical soils in relation to altitude. (*Hardon.*)

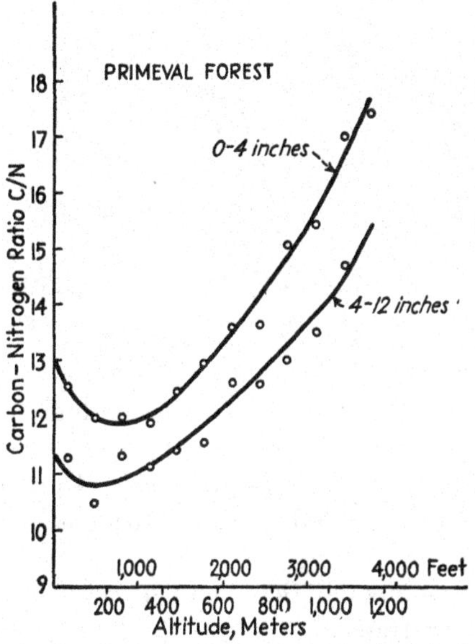

FIG. 91.—Variations of carbon-nitrogen ratio with altitude in the Dutch East Indies. (*Hardon.*)

A_1 horizon is about 6. It declines steadily with elevation to about pH 4 at 5,000 ft. [see Zlatnik and Zvorykin (91)].

Generalized Soil-climate Functions.—When moisture and temperature functions are combined into general climate func-

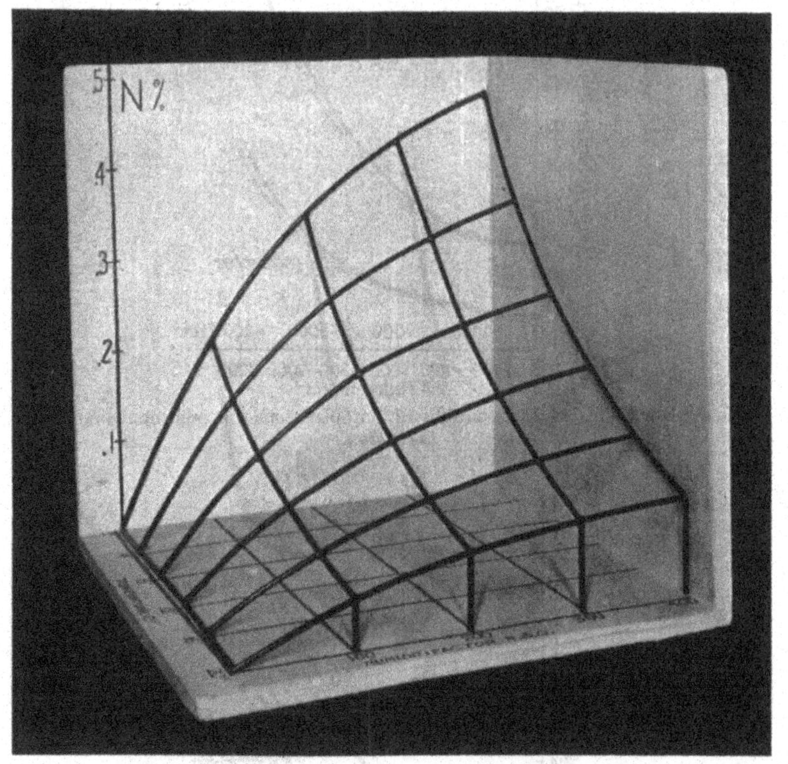

FIG. 92.—Idealized nitrogen-climate surface for the grassland soils of the Great Plains area. Vertical axis—nitrogen; left-right axis—moisture (*NS* quotient); background-front axis—temperature, degrees centigrade.

tions in such a form that the two individual variables are still recognizable, one can visualize much more clearly the variation of soil characteristics over large areas. Combinations of this kind have been constructed for the nitrogen and the clay contents of selected regions.

It will be remembered that the general soil property-climate relationship is written as

$$s = f(m, T)_{o,r,p,t,\ldots} \tag{15}$$

It means that a given soil property s becomes a function of climate when the remaining soil-forming factors are kept constant. Since the equation contains *three* variables, namely, s, m, and T, its graphical representations are surfaces. The two soil-climate surfaces shown in Figs. 92 and 93 were obtained by combining the observed soil-moisture and soil-temperature functions by means of a mathematical technique known as the integration of partial differential equations. In this process it was necessary to make certain assumptions regarding the "constants," and, because of this fact, the surfaces are idealized rather than strictly empirical. Nevertheless, these soil-climate models prove of great value, inasmuch as they enable one to visualize at a glance the approximate major trends of climatic soil formation. Figuratively speaking, such soil-climate functions are bird's-eye views of the soils of entire continents in which the regional features are emphasized at the expense of local details. The latter disadvantage similarly applies to any generalization, such as the Russian transect and Thorp's mountain graph.

The *nitrogen-climate surface* (Fig. 92) may be described by the equation

$$N = 0.55e^{-0.08T}(1 - e^{-0.005m}) \qquad (32)$$

N represents the total nitrogen content of the surface soil, m the moisture expressed as NS quotient, and T the annual air temperature in degrees centigrade. The formula holds only for loamy grassland soils; for clays or sands, correctional terms would be necessary. The equation gives the following information regarding the occurrence of soil nitrogen:

a. When $m = 0$, then also $N = 0$, or, in other words, in desert regions the nitrogen content of the soil tends to be very low, no matter whether the deserts lie in the Arctic or in the tropics.

b. At constant temperature, soil nitrogen increases logarithmically with increasing moisture. The rate of increase depends on temperature and is greatest in the Northern parts and least in the Southern parts of the United States.

c. If moisture is kept constant, soil nitrogen declines exponentially as the temperature rises. The rate of fall depends on moisture: it is greatest in humid and smallest in arid regions. In the United States, southern soils contain less nitrogen and consequently less organic matter than northern soils, for districts of the same moisture level. An idea as to the actual magnitudes of the nitrogen values may be obtained from Table 43.

TABLE 43.—NITROGEN CONTENT OF GRASSLAND SOILS FOR SELECTED MOISTURE AND TEMPERATURE VALUES (GREAT PLAINS AREA)
(Depth from 0 to 7 in.)

	Degrees Centigrade	Increasing annual moisture			
		NS quotient			
		100	200	300	400
Increasing annual temperature	0	0.217% N	0.347% N	0.427% N	0.476% N
	5	0.145% N	0.233% N	0.286% N	0.318% N
	10	0.096% N	0.156% N	0.192% N	0.214% N
	15	0.065% N	0.104% N	0.129% N	0.143% N
	20	0.044% N	0.070% N	0.086% N	0.096% N

The *clay-climate surface* (Fig. 93) is expressed by the equation

$$\Gamma = 0.0114me^{0.140T} \tag{33}$$

Γ denotes the average clay content of the soil profile to a depth

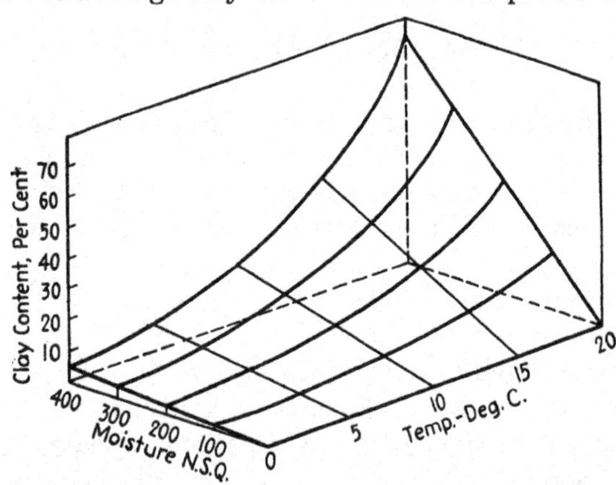

FIG. 93.—Idealized clay-climate surface, showing the variation of "climatic clay" in soils derived from granites and gneisses as a function of moisture and temperature.

of 40 in., m the moisture, and T the temperature. Unlike nitrogen and organic matter, which exist mainly in the surface of the profile and are therefore relatively little influenced by the nature of the parent rock, the clay surfaces are greatly affected by the character of the parent rock. The constants in Eq. (33)

are valid only for soils derived from "mixtures of granites and gneisses." In other words, were the Eastern United States covered uniformly by granites and gneisses, the graph would show the fluctuations of the climatic clay content of the soils formed,

TABLE 44.—CLAY CONTENT (0 TO 40-IN. DEPTH) OF IDEALIZED SOILS DERIVED FROM GRANITES AND GNEISSES FOR SELECTED MOISTURE AND TEMPERATURE VALUES

(Constancy of other soil-forming factors assumed)

	Degrees Centi-grade	Increasing annual moisture			
		NS quotient			
		100	200	300	400
Increasing temperature	0	1.1% clay	2.3% clay	3.4% clay	4.6% clay
	5	2.3% clay	4.6% clay	6.9% clay	9.2% clay
	10	4.6% clay	9.2% clay	13.9% clay	18.5% clay
	15	9.3% clay	18.6% clay	27.9% clay	37.2% clay
	20	18.8% clay	37.5% clay	56.3% clay	(75.0%) clay

neglecting disturbing factors such as glaciation, poor drainage, etc. Some of the more important features of the surface are the following:

a. Under conditions of limited moisture, soils are characterized by paucity of clay (deserts and semideserts).

b. The same holds true for soils with abundant moisture but low temperatures (Arctic and cold regions).

c. At constant temperature the clay content increases with humidity (*NS* quotient). The rate of increase depends on temperature and is greatest in the South and least in the North.

d. Under conditions of constant moisture (*NS* quotient) the clay content augments exponentially with rising temperature. The temperature effect is most pronounced in humid regions and least pronounced in arid ones.

Theoretical clay values for various moisture and temperature figures are listed in Table 44.

A comparison of the nitrogen and the clay surfaces is instructive. Regarding moisture, the effect is essentially similar in both instances, namely, an increase in nitrogen as well as organic matter and clay with ascending *NS* quotient values or rainfall. On the other hand, soils developed in regions of high temperature tend to be high in clay but low in organic matter. In the Northern Hemisphere, soils from the north are likely to be rich in organic matter and poor in clay, whereas southern soils possess much clay but little organic matter.

Existing data on the variation of climate and soil properties with altitude afford a means of verifying the soil-climate surfaces. In Fig. 94, the dotted lines indicate the temperature and NS

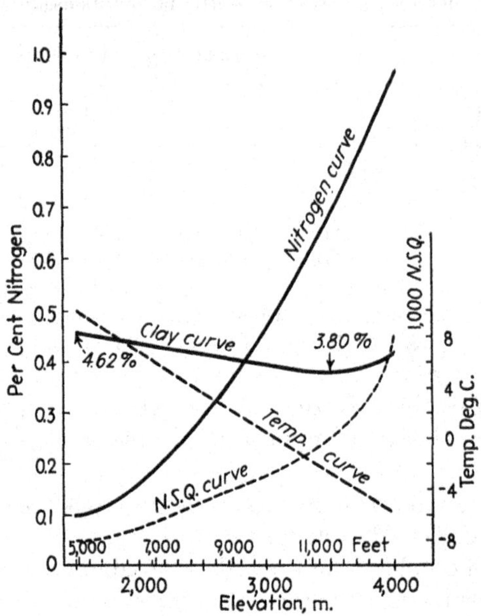

Fig. 94.—Hypothetical curves portraying the variation of nitrogen and clay contents of soils with altitude. Calculations based on Eqs. (32) and (33).

quotient values as a function of altitude in the Rocky Mountains. Their numerical magnitudes roughly correspond to those reported by Hockensmith for Colorado. These climatic figures were inserted into Eqs. (32) and (33) from which the corresponding nitrogen and clay contents were calculated. The solid lines in Fig. 94 mark the trend of soil nitrogen and clay with altitude, on the assumption that the same natural forces that operate in the Great Plains area—where the equations were formulated—also obtain in the mountainous regions. The nitrogen curve climbs exponentially, a result that is in agreement with the work of Hockensmith and Hardon (Figs. 89 and 90). The researches of these investigators therefore speak for the soundness of the nitrogen-climate surface. Interestingly enough, the clay content hardly changes between the elevations 5,000 and 12,000 ft. No experimental data for verifying the clay curve are at hand.

Aggregation of Soils in Relation to Climate.—In the sections on moisture functions, it was shown that the amount of aggregates in soil increases with rainfall. Baver has extended these studies to various parts of the North American continent, and his interesting observations and conclusions are herewith quoted *in extenso* (3).

The extent of aggregation in different soils varies considerably. It is of interest to attempt to correlate aggregate formation with the soil-forming climatic factors, rainfall and temperature. The percentage of aggregates in a given weight of soil is affected considerably by the texture. Coarse-textured soils do not have as much silt and clay to aggregate as the finer textured ones. Consequently, it is of importance not only to know the total percentage of aggregates in the soil but also the amount of silt and clay, in per cent of the total silt and clay present, which is in the form of aggregates.

Aggregate analyses were made on a large number of different soils which have yielded significant data for a correlation between climate and aggregation. Although an insufficient number of samples has been analyzed to formulate accurate mathematical relationships, the results obtained warrant the formulation of several correlations.

The results are summarized graphically in Fig. 95. It is obvious that several striking relationships exist between aggregation and rainfall when temperature is kept constant. The percentage of aggregates in a given weight of soil is a maximum in the semiarid and semihumid regions where the chernozemlike and humid prairie soils are located. Let us discuss the shape of the percentage of aggregates-rainfall curve.

The percentage of aggregates is low in desert soils because of a small clay content. Under arid conditions, chemical weathering does not proceed very far; consequently a small amount of clay is formed from the clay-forming minerals. This small clay content, even though it would all be in the form of secondary particles, would cause a relatively small number of aggregates to be present. As the rainfall becomes greater, chemical weathering is intensified and, according to Fig. 93, clay formation increases. This increase in the clay content of the surface soil obtains until the rainfall becomes great enough to cause eluviation of the clay from the *A* to the *B* horizons. The decrease in the clay content of the *A* horizon diminishes the possibilities for aggregate formation. Not only is the clay content highest in the chernozem and dark humid prairie regions, but the percentage of organic matter in the soils is also the largest. As a result of these two factors the percentage of aggregates in the soil is the greatest in the chernozem and dark humid prairie soils. For example, the amount of aggregates in podsol, dark humid prairie, chernozemlike and chestnut-colored soils, are approximately 10, 40, 40 and 25 per cent, respectively.

If one considers the percentage of silt and clay which is in the form of stable aggregates, the effect of texture becomes insignificant. It is

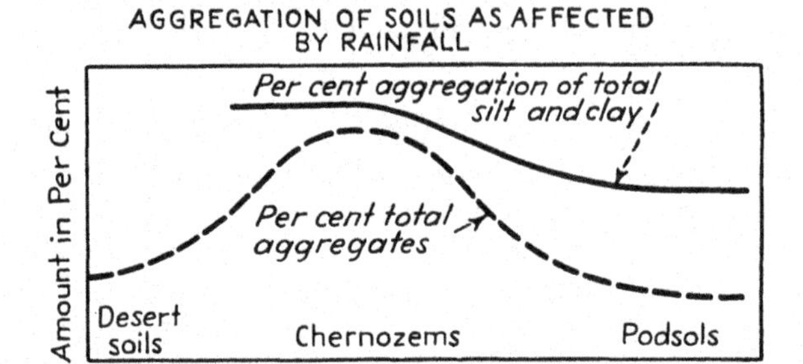

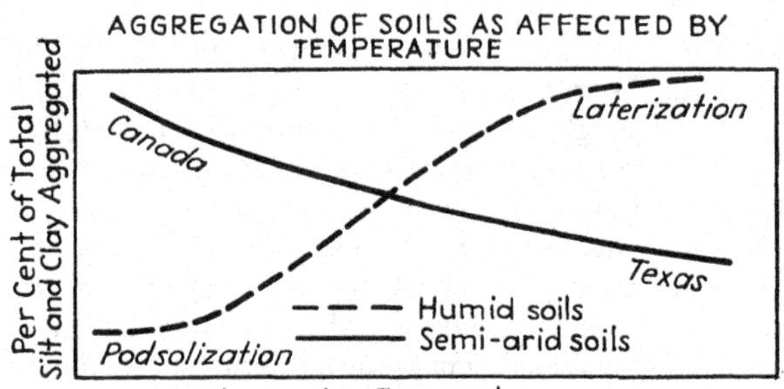

Fig. 95.—Effect of temperature and rainfall on aggregation of soils. (*Baver.*)

seen that podsol soils have a relatively low percentage of the silt and clay aggregated. Only about 25 per cent of the silt and clay present is aggregated. The dark humid prairie, the chernozemlike soils and the chestnut-colored soils have about 50 per cent of the total silt and clay aggregated. The low extent of aggregation in the podsol type of soils is probably due to the removal of alumina and iron from the *A* horizons. The small amount of aggregation is the result of some organic matter which is present in the A_1 horizon. That alumina and iron are responsible for the formation of stable aggregates is evidenced in the extent of aggregation of the coffee-brown layer of the podsol, where about 50 per cent of the silt and clay is aggregated. The large percentage aggregation of the silt and clay in the regions of decreasing rainfall is due to organic matter and bases, especially calcium. The curve seems to flatten at this point, although more data from truly aridic soils would provide information for this group. It should be expected at least to maintain this level if the bases in the soil were calcium and magnesium. If sodium and potassium were present, the percentage aggregation of the silt and clay should be expected to decrease.

When rainfall is kept constant and temperature is increased, the percentage aggregation of the silt and clay is affected differently in humid and semiarid regions. Considering the chernozemlike soils of the semiarid regions the percentage aggregation of the silt and clay decreases from Canada to Texas. Since the calcium content is practically constant, this decrease in aggregation is due to a lowering of the organic-matter content with increasing temperature. The percentage aggregation of these soils varies from 75 per cent in Canada to 25 per cent in Texas.

The humid soils present a somewhat different picture. The percentage aggregation varies from about 25 per cent in the podsols to 95 per cent in the true laterites. The cause associated with the small percentage aggregation of the silt and clay in the podsols has been discussed previously. Aggregation in the laterites is related in some way to the aluminum and iron present. Evidently when the hydrated oxides of these elements dehydrate they cause the formation of exceedingly stable secondary units. Thus, in lateritic soils, alumina and iron contribute to aggregation, while, in the podsol, any secondary particle formation is due to small amounts of organic matter.

Differences in the factors contributing to the building of secondary particles are responsible for the S-shaped curve correlating percentage aggregation of the silt and clay in humid soils with temperature.

Many factors affect aggregate formation, and more study is necessary to correlate and simplify our knowledge of soil structure. This discussion aims at providing a nucleus for further work.

Climatic Cycles and Soil Characteristics.—Climatologists in general are of the opinion that climate is fairly stable and that the average temperature for any consecutive 20 years selected at random from a long record does not differ materially from that of any other consecutive 20 years so selected from that particular record. The climatologist Kincer (43), however, believes that this orthodox conception of the stability of climate needs revision. By means of moving averages he shows that the mean temperature has been rising for at least 40 years in the entire Northeastern United States and southern Canada.

Naturally, this observation presents an additional uncertainty in the evaluation of soil-climate relationships from field studies.

In any given instance, are the present soil-climate relationships really expressions of the effects of today's climate upon the soil? Evidently climatic pulsation cannot be lightly passed over by the student of soil-formation processes.

Evidences of Climatic Cycles (9, 63).—By cycles is meant the reoccurrence of plus and minus departures from the average trend. The cycles are not necessarily periodical. There are short cycles and long cycles, the latter being of particular interest to the soil scientist.

TABLE 45.—SUGGESTED CHRONOLOGY OF POSTGLACIAL CHANGES IN CLIMATE AND VEGETATION (*Sears*, 1932)

Year (past)	Climates	Ohio	Indiana	Iowa
Present	Humid	Mixed forest		Prairie-oak-hickory
1,000		(increasing forest)		(subhumid)
2,000				
3,000	Dry, warm	Oak-hickory-savanna (open country)		Amaranth-grass (semi-arid)
4,000				
5,000				
	Humid	Oak-beech mixed forest (dense forest)		Prairie (subhumid)
6,000				
7,000				
8,000	Dry, cool	Pine-oak	Oak-birch savanna	Amaranth-grass (semi-arid)
		(cool open country)		
9,000				
10,000	Humid, cool	Spruce-pine-fir		Conifers

Short Cycles.—Long-time temperature and rainfall records indicate cycles of a decade or two. Variations in sunspot activities result in 11.5-year cycles. Inspection of growth of *Sequoia gigantea* trees (some of them are three thousand years old) suggest Brückner periods of 34 years duration. Movements of certain modern glaciers reveal an 18-year cycle. Melting of Pleistocene ice sheets as recorded by varved clay deposits indicate periods of from 55 to 70 years and 167 years.

Long Cycles.—The aforementioned clay varves or clay bands with light and dark layers are formed in fresh and slightly brackish water in front of the receding ice border from material brought directly from the melting ice. Thick varves signify warm, clear, and long summers; thin varves mean cold, short, foggy summers. In addition to the short cycles, these clay deposits also point to climatic periods of 1,700 years duration. Peat deposits also have yielded much valuable information on climatic changes. In recent years, the study of pollens (8) held in peat beds has received much attention. Acid peat preserves pollen grains that have been blown into the peat bed during its formation. Identification and counts of the pollen at various

TABLE 46.—NOMENCLATURE AND TIME SCALE OF THE PLEISTOCENE

Glacial periods			Interglacial periods		Duration in years	
	Europe	North America	Europe	North America	Schuchert (71)	Milankovitch (53, 89)
			Present		25,000	21,000
IV	Würm	Wisconsin (Iowan)		(Peorian)	130,000	97,000
			Third (Riss-Würm)	Sangamon	25,000	65,000
III	Riss	Illinoian			45,000	53,000
			Second (Mindel-Riss)	Yarmouth	325,000	193,000
II	Mindel	Kansan			75,000	49,000
			First (Günz-Mindel)	Aftonian	320,000	65,000
I	Günz	Nebraskan			75,000	49,000

depths of the peat permit the construction of pollen diagrams that give the proportion of each kind of grain in each layer. Pollen diagrams show the succession of forest trees surrounding the peat in the course of centuries. In favorable cases, the plant successions record climatic changes and cycles that appear to be of the magnitude of several thousand years. Changes in vegetation as determined by relict indicators, such as prairie areas within deciduous forest, also indicate climatic changes since the retreat of the glaciers. Depopulated areas in arid regions and the succession of alluvial terraces are cited by Huntington (32) as indications of climatic pulsations in historic and prehistoric times.

Extensive climatic cycles on geological scales such as the glacial and interglacial periods are universally recognized. They embrace tens and hundreds of thousands of years (Table 46).

In correlating climatic data with soil properties, we may neglect the cycles of short duration without committing a serious error. Soil formation is a manifestly slow process. Over long periods of time, the relatively short positive and negative departures from the average trend of climate cancel each other. Even with regard to the long cycles we need not be alarmed unduly. It has been estimated (89) that a lowering of a mere 4°C. of the mean annual temperature would readily account for the widespread glaciations in the past. Moreover, in the correlations established between climate and soil, we are primarily interested in gradients rather than in absolute values of moisture and temperature. Thus we are on safe ground to assume that in the Great Plains area the moisture gradient has been in the direction from west to east during the entire period of soil formation. Likewise, the sharp temperature gradient from north to south must have persisted throughout the Pleistocene. Although climatic pulsations must be reckoned with, the soil-climate functions discussed in the preceding sections retain their intrinsic value.

D. DISTRIBUTION OF SOILS ACCORDING TO CLIMATE

The realization of the climatic element in soil formation has led numerous soil scientists to propose systems of soil groupings based on climatic information. Glinka has termed soils that appear to be associated with climate *ectodynamomorphous*, as contrasted with *endodynamomorphous* soils. In the latter group, the climatic habitus is overshadowed by soil characteristics inherent from the parent material or determined by the predominance of some local soil-forming factors.

The essential criterion for a distinct climatic soil type is the presence of similar anatomical and morphological characteristics that are preserved under a variety of geographical environments and geological strata. No matter what the parent material or topography, all soils of a given climatic region must possess certain definite features that are typical for the selected climatic

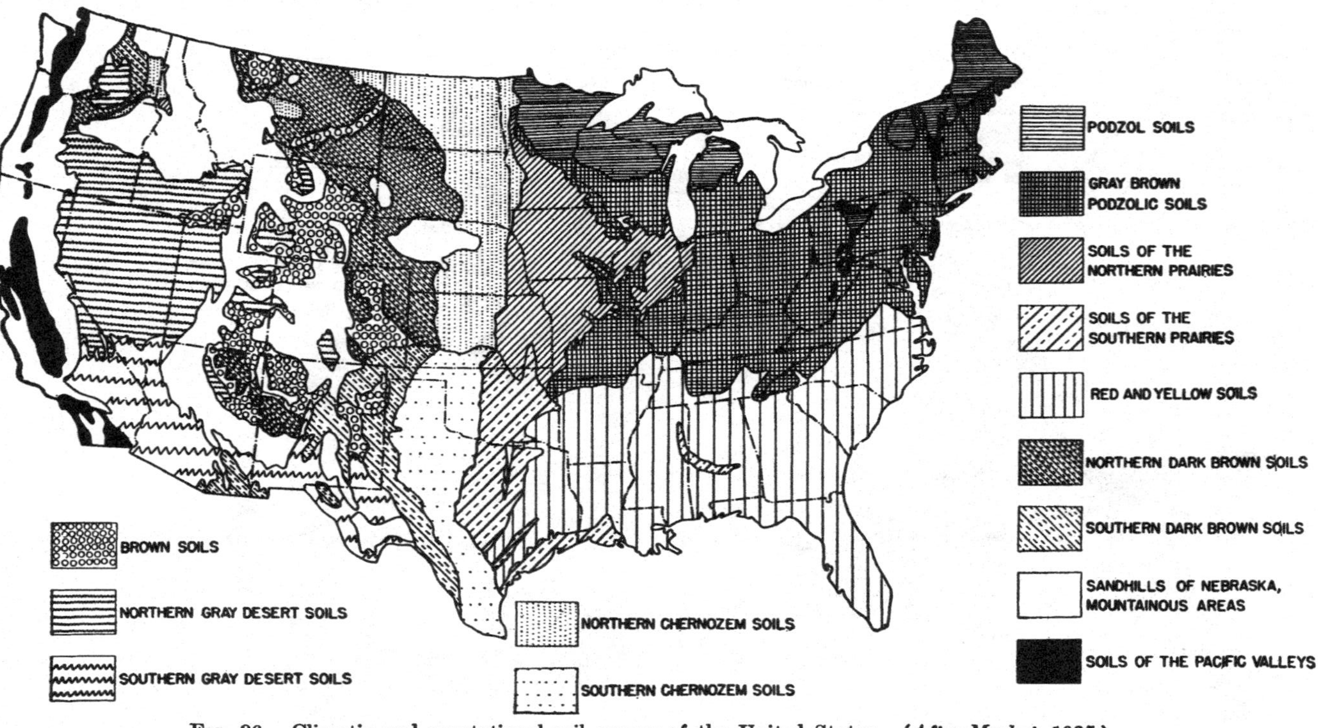

FIG. 96.—Climatic and vegetational soil groups of the United States. (*After Marbut*, 1935.)

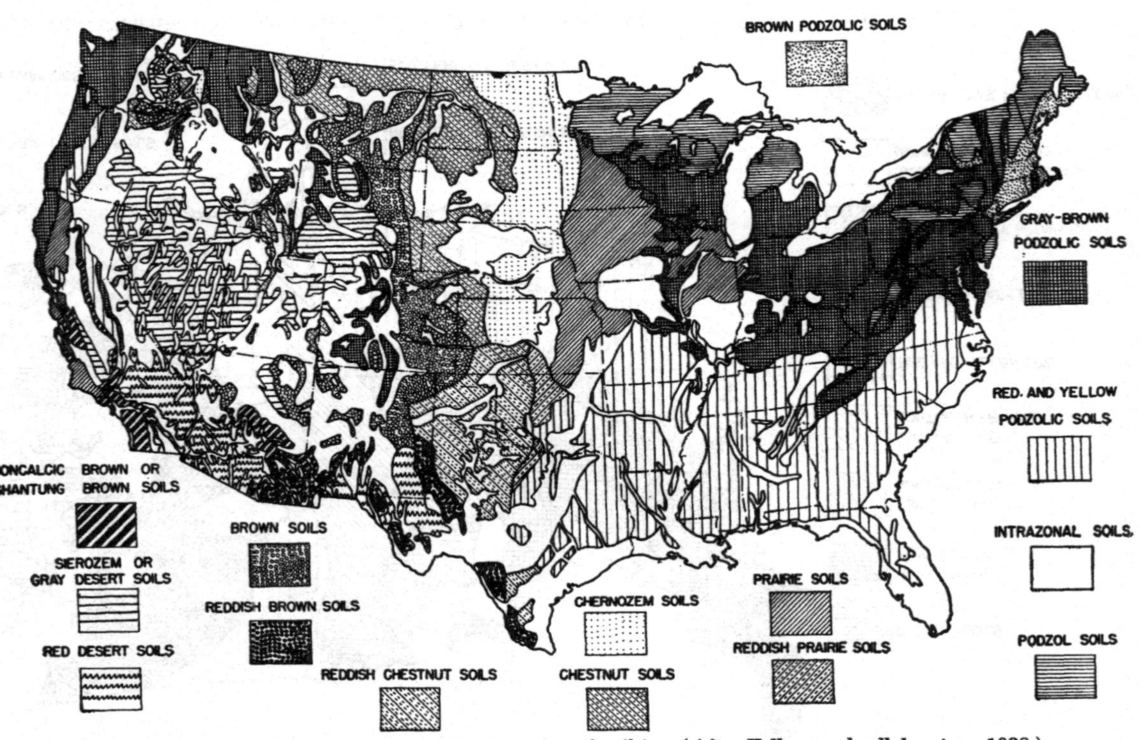

Fig. 97.—Climatic and vegetational soil groups (zonal soils). (*After Kellogg and collaborators*, 1938.)

region. Podsols, for instance, develop in humid cool climates and occur on moraines, alluvial sands, granites, gneisses, diorites, loess, peaty deposits, and even on limestones (rendzina podsols). They are formed on flat topography as well as on slopes, under forest and heath as well as under grassland vegetation.

A sketchlike description of those great soil groups that are often considered to be "climatic soil types," or zonal soils, is given in the following paragraphs;

Skeleton Soils.—Physically disintegrated rock, little chemical weathering: mainly in frigid zones.

Tundra Soils.—Shallow profiles, accumulation of undecomposed plant material, permanent ice in subsoil.

Desert Soils.—Predominance of physical weathering, very low in organic matter, neutral or alkaline reaction. Red desert soils and gray desert soils (Sierozem).

Arid Brown Soils.—Bordering deserts and semideserts; light-brown color, low in organic matter, mostly calcareous.

Chestnut Soils.—Brown or grayish-brown soils of the short-grass region. Considerable organic matter, neutral or alkaline reaction, lime horizon near the surface, indications of columnar structure, little profile development. Also known as dark-brown soils.

Chernozem Soils.—Rich in organic matter to a considerable depth. Neutral to slightly alkaline or acid reaction. Pronounced lime horizon a few feet below the surface, columnar structure, essentially *AC* profile, traces of horizon development.

Steppe Soils.—Term frequently used in European literature. Broad group including chernozems, chestnut soils and arid brown soils.

Prairie Soils.—Intermediate type, bordering chernozems and forest soils. In the United States characteristic of the tall-grass prairies. Rich in organic matter, surface slightly acid, lime horizon at great depth or absent, clay accumulation in subsoils. Some investigators consider the American prairie soils equivalent to the degraded chernozems of Russia. Marbut questions the classification of prairie soils as a climatic soil type.

Gray-brown-podsolic Soils.—Most extensive soil group in the humid temperate part of the United States. Shallow cover of organic matter over a gray-brown horizon of eluviation that rests on a brown *B* horizon.

Brown Forest Soils.—Wide distribution in western Europe. Moderate development of *A* and *B* horizons. Leached, slightly acid, brown to dark brown depending on organic matter and parent material. No carbonate horizon. Essentially the same as Ramann's brown earth: sometimes called podsolized soils or podsolic soils. Unlike many European workers, Kellogg considers the brown forest soils to be essentially nonclimatic or intrazonal soils.

Podsol Soils.—Characterized by pronounced *A*, *B*, and *C* horizons. Surface rich in organic matter, followed below by a white or ash-gray leached horizon that is above the brown zone of accumulation of aluminum and iron. Feebly developed podsols are often designated as podsolized soils.

Yellow and Red Soils.—Soils of humid warm regions. Low in organic matter, strongly leached, rich in clay, bright yellow and red colors. Many varieties and subgroups.

Laterite Soils.—Advanced stage of rock decomposition. Silica leached out, accumulation of sesquioxides in the surface, iron crusts, hardpans. Classification still controversial.

For more detailed information on climatic soil types, the reader is referred to the books of Joffe (40), Robinson (64), and the U. S. Department of Agriculture *Yearbook*, 1938.

Systems of Climatic Soil Classification.—Lang's (46) climatic grouping of soils is of historical interest. Soils are arranged according to rain factors. Temperature is considered insofar as it influences evaporation, but little attempt is made to recognize its direct effect on soil formation. On the basis of Lang's system, the calcareous chestnut soils of the Dakotas should belong to the same class as the acid, reddish, lateritic soils of Alabama. Further discrepancies may be found by comparing the rain-factor map with any soil map of the United States.

Meyer (52) presents a classification of soils based on the *NS* quotient (Table 47). In principle he, too, ignores the temperature effect, but in practice he seems to realize that the efficiency of the *NS* quotient is not the same in all latitudes. Groupings such as "north Russian region" or "Mediterranean region" are essentially temperature classifications.

Attention should be called to a paper by Shostakovich (73) who classifies the climatic soil types of the world according to the average humidity of the soil-forming period (vegetational period from spring to fall) and the average temperature of the same period. It corresponds essentially to the rain factor of the vegetational period. The reason for this procedure rests on the idea that in regions with cold winters the soil processes occur mainly during the frost-free season. Similar considerations have led Reifenberg (61) to classify the soils of Palestine according to the rain-factor sum of the winter months, because the summers are so dry that no chemical processes take place.

TABLE 47.—MEYER'S CLASSIFICATION OF SOILS ACCORDING TO *NS* QUOTIENTS (EUROPE)

Soil Zones and Climatic Zones	*NS* Quotient (Annual Values)
Deserts and desert steppes	0– 100
Mediterranean region	50– 200
Chestnut soils	100– 275
Chernozems	125– 350
(Brown earth)	275– 400
Atlantic regions	375–1,000
Heath	375– 700
North Germanic Scandinavian region	300–1,200
North Russian region	400– 600
Tundras	Over 400
High mountains	1,000–4,000

Hesselman (27) has applied de Martonne's aridity index to the distribution of soils and vegetation in Sweden and found satisfactory agreement. It is of interest to note that Hesselman is compelled to establish temperature subgroups whenever the moisture belt transgresses several degrees of latitude. A comparison between the aridity index and the ordinary rainfall map convincingly shows that the latter completely fails to portray soil-climate relationships. There exists the paradoxical fact that in southern Sweden the slightly acid brown forest soils are formed under 27 in. of rainfall, whereas the true podsols with a thick coating of highly acidic raw humus require but 18 in. of precipitation. In comparison, the aridity indexes are 30 for the brown forest soil and 50 for the podsols, which is in better agreement with the facts.

In certain respects, the systems of Lang and of Meyer are less satisfactory than the old classification system of Ramann. This author arranged the soil types along two coordinates, humidity and temperature, both of them receiving equal consideration. Independently, Vilensky developed a similar but more detailed

TABLE 48.—VILENSKY'S CLIMATIC SYSTEM OF SOILS

Heat zones	Temperature ↓ / Moisture →	Humidity zones (precipitation-evaporation ratio)				
		0-0.25	0.25-0.75	0.75-1.25	1.25-1.75	1.75-2.25
		Arid	Semiarid	Feebly arid	Semihumid	Humid
Polar	°C. −12 to −4	Tundra soils	Semibog soils	Bog soils	?	Concealed podsols; podsolized bog soils
Cold	− 4 to +4	Sward soils	?	Black meadow soils	Degraded meadow soils	Podsolized soils
Temperate	+ 4 to 12	Gray soils	Chestnut brown soils	Chernozem (black soils)	Degraded soils (gray forest soils)	Podsolized soils
Subtropic	12 to 20	?	Yellow soils of arid steppe	Yellow soils	Degraded yellow soils	Podsolized yellow soils
Tropic	Above 20	Red soils of semideserts	Red soils	Laterite soils	Degraded red soils	Podsolized red soils

TABLE 49.—THORNTHWAITE'S CLIMATIC SYSTEM OF SOILS

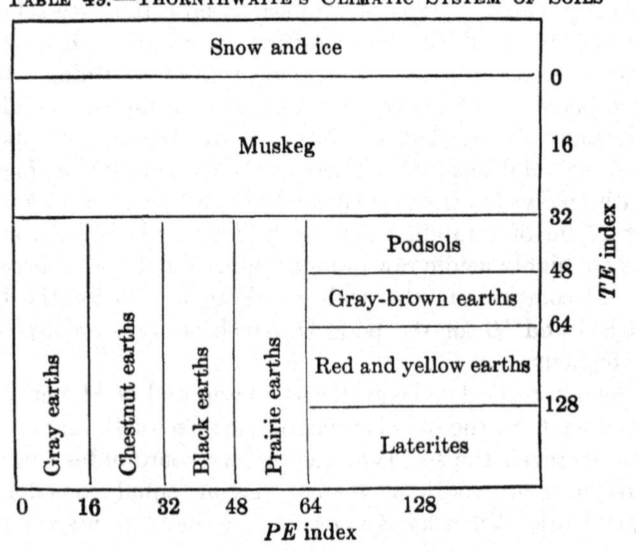

system. Moreover, the variables are given numerical values (Table 48).

Vilensky's (85) system presents the most comprehensive climatic soil classification available, and a careful study of Table 48 will repay the reader. However, certain features must be accepted with reservation. For instance, the chernozems and the laterites appear in the same vertical column, and one would have to conclude that the profound differences in soil properties are caused by a mere temperature shift of 10 or 20°C.

Thornthwaite (77) has given special attention to the climatic classification of the soils of North America as based on precipitation effectiveness and temperature efficiency. According to Table 49 moisture dominates soil development in the grassland areas, whereas in timbered regions temperature plays the decisive role.

Assuming that the climatic classification of soil is basically sound, one would expect that regions of similar soil profiles in America and Europe would be characterized by similar climates. On the basis of the general soil map of Europe and a tentative American soil map by Marbut, Jenny has examined the existing information and arrived at the conclusions summarized in Table 50. The consistency of climatic data for European and American podsols, chernozems, desert soils, brown forest soils and gray-brown-podsolic soils is rather satisfactory.

Prescott (58) has undertaken a similar study in Australia. After the major soil zones had been mapped from field observations, he proceeded to correlate soil distribution with climatic data. Since annual temperatures do not vary excessively (60 to 80°F.), Prescott restricted the investigation to moisture factor relationships, mainly rain factor and *NS* quotient. Generally speaking, it appears that under the high temperatures of Australia all climatic soil boundaries have been shifted to lower *NS* quotient values as compared with the temperate zone of North America and Europe. This is especially striking for the black and brown earths and the podsols. According to Table 50, the chernozems have an *NS* quotient range from 130 to 250 and higher, whereas in Australia black earths are found between *NS* quotients 50 and 200. The climatic boundaries overlap considerably, and Prescott concludes "that the *NS* quotient is not adequate to express the limits of any particular soil type." He

TABLE 50.—PARALLELISM OF SOIL-CLIMATE RELATIONSHIPS IN EUROPE AND THE UNITED STATES
(Temperate regions only, 4–12°C., 39–54°F.)

Climatic soil types	Moisture factor (annual *NS* quotient)	
	Europe	U. S. A.
Gray desert soils	100	30–110
Arid brown soils	....	60–120
Chestnut soils	140– 270	100–180
Chernozems	130– 250 (350)	140–250 (320)
Arid-humid boundary	200	200–250
Degraded chernozem	250– 350	
Prairie soils	....	260–350 (420)
Brown forest soils and gray-brown-podsolic soils	320– 460	280–400
Podsols	400–1,000	380–750

NOTE: The figures in parentheses denote extreme values.

FIG. 98.—Chernozem soil near Kharkov, Ukraina, U.S.S.R. (*From the collection of photographs of the late Prof. C. F. Shaw.*)

believes that the character of the rainfall also should be considered. "In Australia only in regions of summer rainfall are soils akin to the true chernozem produced, whereas in the zone of winter rainfall the only black soils developing are probably akin to the rendzina or lime-humus soils." Before we cast final judgment on the general validity of the climatic systems of soil classification, let us consider briefly the interplay of climate with other soil-forming factors.

Rivalry between Climatic and Other Soil-forming Factors.— The climatic boundaries between soil types are seldom sharply defined. They represent fringes rather than definite lines. As an example, one might refer to Table 50. The chestnut soils in America range from *NS* quotient 100 to 180: the chernozems range from 140 to 250, an overlapping of 40 *NS* quotient units. Scherf (69), who has made a detailed study of the *NS* quotient-soil relation in certain European countries, offers an explanation that throws light on the limitations of the climatic concept of soil classification. Scherf is particularly concerned with the European climatic sequence;

Chernozem	→ Brown forest soils →	Podsols
(arid-humid boundary)	(semihumid and humid)	(humid)

He observed by extensive field studies that the *water permeability* of the subsoil and its *carbonate content* greatly modify the climatic boundaries. Under *NS* quotient 400 to 500, podsols are formed on permeable sandy soils, whereas on calcareous shales brown forest soils result. Similarly, in areas of *NS* quotient 275 to 350, chernozems are observed on loams and clay loams, whereas brown forest soils develop on the lighter sandy loams. Evidently, in the transition zones the velocity of the vertical water flow may be responsible for the fringed boundaries. For central European conditions, Scherf sets up the following instructive scheme:

NS quotient

200–275 *Climatic chernozem.* The influence of climate is greater than that of the parent material. Chernozems are formed on light as well as on heavy-textured soils.

275–350 *Transition from chernozem to brown forest soils.* The parent material controls soil development. Low water permeability gives chernozem, high water permeability leads to brown forest soils.

350–500 *Climatic brown forest soils.* Climate predominates over parent material.

375–500 *Transition from brown forest soils to podsols.* Endogenic factors govern the establishment of soil types. Brown forest soils occur in poorly drained areas, podsols on well-drained parent materials.

500–600 *Climatic podsols.* Any rock will develop into a podsol. The climate overshadows the influence of other soil-forming factors.

Inherent Limitations of the Climatic Systems of Soil Classification.—In view of the fundamental equation of soil-forming factors

$$s = f(cl, o, r, p, t, \cdots) \tag{4}$$

the difficulties that confront climatic systems of soil classification are at once apparent. We cannot completely describe a multivariable system on the basis of a single variable. Climate could singularly impress its features on the soils of the world only if the surface of the earth were covered with a uniform parent material and if it had the same topographic features, similar types of organisms, and comparable time factors.

The success of climatic systems of soil classification is in considerable measure due to the fact that in many parts of the world several of the conditioning variables are reasonably constant as compared with variations in climatic factors. This is especially true for certain parts of Russia and for the extensive American plateau between the Rocky Mountains and the Appalachians. In regions such as the extreme Western part of the United States where parent materials form a mosaic pattern of sands, granites, peats, quartzites, marine clays, serpentines, limestones, etc., climatic systems of soil classification are of limited value.

In the opinion of the author, a step in the right direction has been undertaken by Marbut (47). In his classification of soils into pedocals and pedalfers, or soils with and without a lime horizon, he realized that this soil property is a true reflection of environment only if the remaining soil formers are controlled. The importance of this restriction is immediately evident from the following formulation:

$$s = f\,(\text{climate})_{o,r,p,t,\ldots} \tag{34}$$

Although Marbut does not actually state the relationship indicated by Eq. (34), there is no question that he saw the problem clearly. Suppose we examine a profile in the field and note the absence of a lime horizon. Is this soil a pedalfer? Not necessarily, according to Marbut. A number of restrictions must be observed. First, the soil must be virgin, *i.e.*, the lime horizons must not have been eliminated artificially by human interference (*e.g.*, by irrigation practices). Second, the soil must be associated with what Marbut calls "normal" topography, *i.e.*, a slightly rolling land surface; soils on flat areas or steep slopes must be excluded. Third, the parent material must be such as

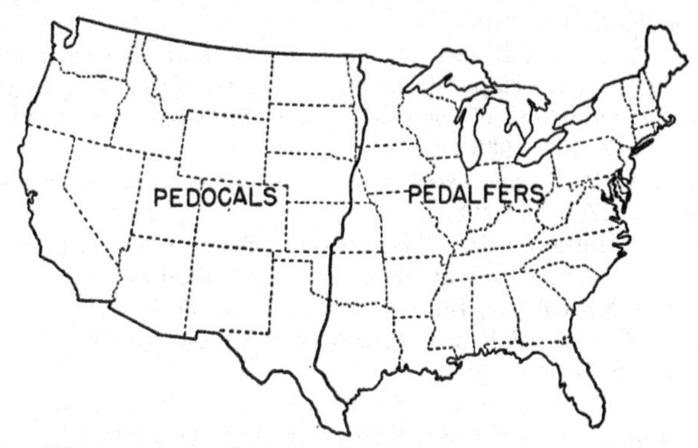

FIG. 99.—Marbut's classification of soils into pedocals and pedalfers.

to be able to produce earth alkali carbonates. Fourth, and most important of all, the profile must be mature. Now, it may readily be seen that these special conditions are embodied in Eq. (34). The restrictions imposed merely assign special values to r, p, and t.

In Fig. 99, not all soils that are embraced by the area labeled as pedocals contain lime horizons. Only those soils which have the required normal topography and which have attained sufficient maturity deserve the attribute pedocalic. Corresponding limitations obtain in the region of the pedalfers.

In spite of inherent limitations, the climatic or zonal systems of soil classifications are appealing and will probably continue to enjoy popularity, because they emphasize the profoundly important variables, moisture and temperature, which not only influence soils but intimately affect the actions and life of man.

Literature Cited

1. ALWAY, D. F.: The loess soils of the Nebraska portion of the transition region, I, II, III, *Soil Sci.*, **1**:197–258, 299–316, 1916.

2. Angström, A.: A coefficient of humidity of general applicability, *Geog. Annal.*, **18**:245–254, 1936.
3. Baver, L. D.: A classification of soil structure and its relation to the main soil groups, *Am. Soil Survey Assoc. Bull.* 15: 107–109, 1934.
4. Baver, L. D., and Scarseth, S. D.: Subtropical weathering in Alabama as evidenced in the Susquehanna fine sandy loam profile, *Soil Research*, **2**:288–307, 1930.
5. Bennett, H. H., and Allison, R. V.: The soils of Cuba, Trop. Plant Research Found., Washington, D. C., 1928.
6. Brown, I. C., and Byers, H. G.: The chemical and physical properties of dry-land soils and of their colloids, U. S. Department of Agriculture, *Tech. Bull.* 502, 1935.
7. Brown, I. C., and Byers, H. G.: Chemical and physical properties of certain soils developed from granitic materials in New England and the Piedmont and of their colloids, U. S. Department of Agriculture, *Tech. Bull.* 609, 1938.
8. Cain, S. A.: Pollen analysis as a paleo-ecological research method, *Botan. Rev.*, **5**:627–654, 1939.
9. Climatic Variations, *Bull. Am. Meteorol. Soc.*, 19, No. 5, 1938.
10. Coffey, G. N.: A study of the soils of the United States, U. S. Department of Agriculture, Bureau of Soils, *Bull.* 85, 1912.
11. Corbet, A. S.: "Biological Processes in Tropical Soils," W. Heffer & Sons, Ltd. London, 1935.
12. Craig, N.: Some properties of the sugar cane soils of Mauritius, Mauritius Department of Agriculture, *Bull.* 4: 1–35, 1934.
13. Craig, N., and Halais, P.: The influence of maturity and rainfall on the properties of Lateritic soils in Mauritius, *Empire J. Exp. Agr.*, **2**: 349–358, 1934.
14. Crowther, E. M.: The relationship of climatic and geological factors to the composition of soil clay and the distribution of soil types, *Proc. Roy. Soc.*, B 107: 1–30, 1930.
15. Davis, F. L.: A study of the uniformity of soil types and of the fundamental differences between the different soil series, *Alabama Agr. Expt. Sta. Bull.* 244, 1936.
16. Dean, A. L.: Nitrogen and organic matter in Hawaiian pineapple soils, *Soil Sci.*, **30**:439–442, 1930.
17. Feustel, I. C., Dutilly, A., and Anderson, M. S.: Properties of soils from North American Arctic regions, *Soil Sci.*, **48**:183–198, 1939.
18. Filatov, M. M.: Diagrammatic meridional soil profile of the European part of U.S.S.R., Appendix to the journal *Pedology*, 2, Moscow, 1927.
19. Fox, C. S.: "Bauxite and Aluminous laterite," London, 1932.
20. Fox, C. S.: Buchanan's laterite of Malabar and Kanara, *Records Geol. Survey India*, **69**:389–422, 1936.
21. Gedroiz, K. K.: Der adsorbierende Bodenkomplex und die adsorbierten Bodenkationen als Grundlage der genetischen Bodenklassifikation, *Kolloid chem. Beihefte*, **29**:149–260, 1929.
22. Gillam, W. S.: The geographical distribution of soil black pigment, *J. Am. Soc. Agron.*, **31**:371–387, 1939.

23. GLINKA, K.: "Die Typen der Bodenbildung, ihre Klassifikation und geographische Verbreitung," Verlagsbuchhandlung Gebrüder Borntraeger, Berlin, 1914.
24. GRIFFITH, B. G., HARTWELL, E. W., and SHAW, T. E.: The evolution of soils as affected by the old field white pine—mixed hardwood succession in central New England, *Harvard Forest Bull.* 15, 1930.
25. HARDON, H. J.: Factoren die het organische stof-en het stikstof-gehalte van tropische gronden beheerschen, *Medeleelingen alg. Proefst. Landbouw.*, 18, Buitenzorg, 1936.
26. HARRASSOWITZ, H.: Laterit, *Fortschr. Geologie und Paleont.*, **4**:253–566, 1926.
27. HESSELMAN, H.: Die Humidität des Klimas Schwedens und ihre Einwirkung auf Boden, Vegetation und Wald, *Medd. Statens Skogsförsöksanstalt*, **26**, 4, 515–559, 1932.
28. HILGARD, E. W.: "Soils," The Macmillan Company, New York, 1914.
29. HIRTH, P.: Die Isonotiden, *Ref. Peterm. Mitt.*, **72**:145–149, 1926.
30. HOCKENSMITH, R. D., and TUCKER, E.: The relation of elevation to the nitrogen content of grassland and forest soils in the Rocky Mountains of Colorado. *Soil Sci.*, **36**:41–45, 1933.
31. HOSKING, J. S.: The ratio of precipitation to saturation deficiency of the atmosphere in India, *Current Sci.*, **5**:422, 1937.
32. HUNTINGTON, ELLSWORTH: Climatic pulsations, *Geog. Annal.*, Sven Hedin volume: 571–608, 1935.
33. ISAAC, W. E., and GERSHILL, B.: The organic matter content and carbon-nitrogen ratios of some semi-arid soils of the Cape Province, *Trans. Roy. Soc. South Africa*, **23**:245–254, 1935.
34. JENNY, H.: Klima und Klimabodentypen in Europa und in den Vereinigten Staaten von Nordamerika, *Soil Research*, **1**:139–187, 1929.
35. JENNY, H.: A study on the influence of climate upon the nitrogen and organic matter content of the soil, *Missouri Agr. Expt. Sta. Research Bull.* 152, 1930.
36. JENNY, H.: Behavior of potassium and sodium during the process of soil formation, *Missouri Agr. Expt. Sta. Research Bull.* 162, 1931.
37. JENNY, H.: The clay content of the soil as related to climatic factors, particularly temperature, *Soil Sci.*, **40**:111–128, 1935.
38. JENNY, H.: "Properties of Colloids," Stanford University Press, Stanford University, Calif., 1938.
39. JENNY, H., and LEONARD, C. D.: Functional relationships between soil properties and rainfall, *Soil Sci.*, **38**:363–381, 1934.
40. JOFFE, J. S.: "Pedology," Rutgers University Press, New Brunswick, N. J., 1936.
41. KELLEY, W. P., and BROWN, S. M.: Replaceable bases in soils, *California Agr. Expt. Sta. Tech. Paper* 15, 1924.
42. KELLEY, W. P., JENNY, H., and BROWN, S. M.: Hydration of minerals and soil colloids in relation to crystal structure, *Soil Sci.*, **41**:259–274, 1936.
43. KINCER, J. B.: Is our climate changing? A study of long-time temperature trends, *Monthly Weather Rev.*, **61**:251–259, 1933.

44. Kincer, J. B.: Precipitation and humidity, *Atlas of American Agriculture*, Part II, Washington, D. C., 1922.
45. Lacroix, H.: Les latérites de la Guinée et les produits d'altération qui leur sont associés, *Nouvelles Archives du Muséum d'Historie Naturelle, Ser.*, 5, **5**:255–358, 1913.
46. Lang, R.: "Verwitterung und Bodenbildung als Einführung in die Bodenkunde," E. Schweizerbart'sche Verlagsbuchhandlung, Stuttgart, 1920.
47. Marbut, C. F.: A scheme for soil classification, *Proc. First Intern. Congr. Soil Sci.*, **4**, 1–31, 1928.
48. Marbut, C. F.: The relation of soil type to organic matter, *J. Am. Soc. Agron.*, **21**:943–950, 1929.
49. Marbut, C. F.: Soils of the United States, *Atlas of American Agriculture*, Part III, Washington, D. C., 1935.
50. Martonne, E. de: Aréism et indice d'aridité, *Compt. rend.*, **182**:1395–1398, 1926.
51. Mattson, S., and Gustafsson, Y.: The chemical characteristics of soil profiles, I, The Podsol, *Lantbruks-Högskol. Ann.*, I: 33–68, 1934.
52. Meyer, A.: Über einige Zusammerhänge zwischen Klima und Boedn in Europa, *Chemie der Erde*, **2**:209–347, 1926.
53. Milankovitch, M.: "Théorie mathématique des phénomènes thermiques produits par la radiation solaire," Gauthier-Villars & Cie, Paris, 1920. See also: Köppen, W. and Wegener, A.: "Die Klimate der geologischen Vorzeit," Verlagsbuchhandlung Gebrüder Borntraeger, Berlin, 1924.
54. Miller, M. F., and Krusekopf, H. H.: The influence of systems of cropping and methods of culture on surface runoff and soil erosion, *Missouri Agr. Expt. Sta. Research Bull.* 177, 1932.
55. Mohr, E. C. J.: "De Grond van Java en Sumatra," J. H. de Bussy, Amsterdam, 1922.
56. Penck, A.: Versuch einer Klimaklassifikation auf physiogeographischer Grundlage, *Sitzber. preuss. Akad. Wiss. physik. math. Klasse, Berlin*, 236, 1910.
57. Pendleton, R. L.: Papers presented at the Sixth Pacific Science Congress, Berkeley, California, 1939 (in print). See also: *Am. Soil Survey Assoc. Bull.* 17: 102–108, 1936.
58. Prescott, J. A.: The soils of Australia in relation to vegetation and climate. Council for Scientific and Industrial Research, Australia, *Bull.* 52, 1931.
59. Prescott, J. A.: Single value climatic factors, *Trans. Roy. Soc. South Australia*, **58**:48–61, 1934.
60. Ramann, E.: "Bodenkunde," Verlag Julius Springer, Berlin, 1911.
61. Reifenberg, A.: Die Bodenbildung im südlichen Palestina in ihrer Beziehung zu den klimatischen Faktoren des Landes, *Chemie der Erde*, **3**:1–27, 1927.
62. Reifenberg, A.: Die Zusammensetzung der Kolloidfraktion des Bodens als Grundlage einer Bodenklassifikation, *Verhandl. der zweiten Kom-*

mission und der Alkali-Subkommission der Int. Bodenkundl. Gesellschaft, Teil A:141–153, Kopenhagen, 1933.
63. Reports of the conferences on cycles, Carnegie Institution, 1929.
64. ROBINSON, G. W.: "Soils. Their Origin, Constitution and Classification," Thomas Murby, London, 1932.
65. ROBINSON, W. O., and HOLMES, R. S.: The Chemical composition of soil colloids, U. S. Department of Agriculture, *Bull.* 1311, 1924.
66. RUSSELL, J. C., and ENGLE, E. G.: Soil horizons in the central prairies. *Rept. Fifth Meeting Am. Soil Survey Assoc.,* **6**:1–18, 1925.
67. RUSSELL, J. C., and McRUER, W.: The relation of organic matter and nitrogen content to series and type in virgin grassland soils, *Soil Sci.,* **24**:421–452, 1927.
68. SAUSSURE, H. B. DE: "Voyages dans les Alpes," **5**, 208, 1796.
69. SCHERF, E.: Über die Rivalität der boden und luftklimatischen Faktoren bei der Bodenbildung, *Annal. Inst. Regii Hungarici Geol.,* **29**:1-87, 1930.
70. SCHUBERT, A.: Das Verhalten des Bodens gegen Wärme, *Blanck's Handbuch der Bodenlehre,* **6**:342–375, 1930.
71. SCHUCHERT, C.: "Outlines of Historical Geology," John Wiley & Sons, Inc., New York, 1931.
72. SEARS, P. B.: The archeology of environment in eastern North America, *Am. Anthropologist,* **34**:601–622, 1932.
73. SHOSTAKOVICH, V. B.: Versuch einer Zusammenstellung der Bodenformationen mit den Klimabedingungen, *Proc. Second Intern. Congr. Soil Sci.,* **5**:215–222, Moscow, 1932.
74. SIEVERS, F. J., and HOLTZ, H. F.: The influence of precipitation on soil composition and on soil organic matter maintenance, *Washington Agr. Expt. Sta. Bull.* 176, 1923.
75. SIGMOND, A. A. J. DE: "The Principles of Soil Science," Thomas Murby Company, London, 1938.
76. Soils of the United States, U. S. Department of Agriculture, *Yearbook,* **1938**:1019–1161, Washington, D. C., 1938.
77. THORNTHWAITE, C. W.: The climates of North America, *Geog. Rev.,* **21**: 633–654, 1931.
78. THORP, J.: The effects of vegetation and climate upon soil profiles in northern and northwestern Wyoming, *Soil Sci.,* **32**:283–301, 1931.
79. THORP, J.: Soil profile studies as an aid to understanding recent geology. *Bull. Geol. Soc. China,* **14**:359–392, 1935.
80. TRANSEAU, E. N.: Forest centers of eastern America, *Am. Naturalist,* **39**:875–889, 1905.
81. U. S. Department of Agriculture, Weather Bureau, *Bull.* W., ed. 2, 1926.
82. VAGELER, P.: "Grundriss der tropischen und subtropischen Bodenkunde," Verlagsgesellschaft für Ackerbau, Berlin, 1930.
83. VAN'T HOFF, J. H.: "Études de dynamique chimique," Amsterdam, 1884.
84. VEIHMEYER, F. J.: Some factors affecting the irrigation requirements of deciduous orchards, *Hilgardia,* **2**:125–284, 1927.
85. VILENSKY, D. G.: Concerning the principles of genetic soil classification. *Contributions to the study of the soils of Ukraina,* **6**:129–151, Kharkow, U.S.S.R., 1927.

86. Waksman, S. A., and Gerretsen, F. C.: Influence of temperature and moisture upon the nature and extent of decomposition of plant residues by microorganisms. *Ecology*, **12**:33–60, 1931.

87. Weis, F.: Physical and chemical investigations on Danish Heath soils (Podsols), *Kgl. Danske Videnskab. Selskab. Biol. Medd.*, VII, 9, 1929.

88. Wiegner, G.: "Boden und Bodenbildung," T. Steinkopff, Leipzig, 1924.

89. Woldstedt, P.: "Das Eiszeitalter," F. Enke, Stuttgart, 1929.

90. Wollny, E.: "Die Zersetzung der organischen Stoffe und die Humusbildungen," Heidelberg, 1897.

91. Zlatnik, A., and Zvorykin, I.: Essai des recherches du changement périodique de la station forestière et celle des prairies, *Bull. inst. Nat. Agron.*, Brno, R.C.S., Sign. D 19, 1932.

CHAPTER VII

ORGANISMS AS A SOIL-FORMING FACTOR

Soil scientists do not agree among themselves as to the exact place of organisms in the scheme of soil-forming factors. Nikiforoff, Marbut, and others contend that life in general and vegetation in particular are the most important soil formers. "Without plants, no soil can form," writes Joffe in his "Pedology." On the other hand, Robinson, in his discussion of the soils of Great Britain writes:

> Vegetation cannot be accorded the rank of an independent variable, since it is itself closely governed by situation, soil, and climate. And, therefore, whilst the intimate relationship between natural vegetation and soil cannot be overlooked, it must be regarded as mainly a reciprocal contract.

Likewise, in all studies of soil-climate relationships, vegetation is treated as a dependent variable rather than as a soil-forming factor, because significant changes in climate are always accompanied by variations in kind and amount of plant life. In the ensuing sections, we shall attempt to clarify this controversy and to elucidate the exact role of organisms as soil-forming factors.

A. DEPENDENT AND INDEPENDENT NATURE OF ORGANISMS

It is a universally known truism that microorganisms, plants, and many higher animals affect and influence the properties of soil. But such action alone in no way establishes organisms as soil-forming factors. The mere "acting" is neither a sufficient nor essential part of the character of a soil-forming factor. The *A* horizon of a given soil acts upon the *B* horizon, and yet obviously it is not a soil-forming factor. *If organisms are to be included among the soil formers, they must possess the properties of an independent variable* as indicated in the introductory chapter. It must be shown that the factor organisms can be made to change in its essential characteristics, while all other soil-forming factors such as climate, parent material, topography, and time are maintained at any particular constellation. This variation must be attainable under experimental conditions or by appropriate selections in the field.

For the successful comprehension of the controversial and complicated situation, let us resort to an imaginary laboratory experiment. A number of soils, S_1, S_2, S_3, etc., that differ widely in their properties are sterilized and inoculated with a population of microorganisms consisting of a large number of varied species. This group of organisms represents a biological complex and is given the symbol B. After a period of days or weeks, we shall observe that each soil contains a specific type of microbiological population depending on the properties of the soil and the conditions of incubation. These resultant populations or biological complexes in the different soils we shall designate with the symbols b_1, b_2, b_3, etc. The relationships between the initial and the resulting biological complexes may be represented as follows:

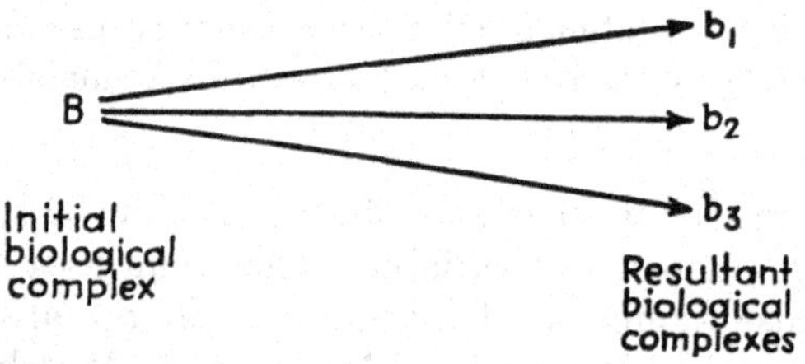

The initial biological complex B deserves the status of an independent variable because, obviously, we may add any kind of biological complex to our group of soils, regardless of their properties. The initial complex B, therefore, assumes the role of a soil-forming factor. An entirely different situation exists for the resultant biological complexes b_1, b_2, b_3. Here both the species of microorganisms that survive and the amounts of each (yields) are functions of the soils and of their soil-forming factors. These resultant complexes are dependent variables, in other words, they are not soil-forming factors, though, of course, they influence the properties of the soil.

The foregoing conclusions need not be restricted to the microbiological population of the soil. They apply to all types of organisms, microbes, vegetation, animal life, and, to some extent, to man. For, in principle, we may conduct our imaginary laboratory experiment with any group of organisms.

In view of the properties of an independent variable, which is defined as a variable that can be made to vary independently of other variables, we come to realize at once that the quantity or yield of a given group of organisms can never be a soil-forming factor, because it is completely governed by the properties of the soil and the environment. In speaking of plants, animals, etc., as soil formers, we must modify our habitual picture and divorce the quantity aspect from the concept of organisms. We must focus our attention on the quality aspect, *i.e.*, the kind and relative frequency of species.

The Biotic Factor.—This quality aspect of organic life constitutes the essential part of the *biotic factor of soil formation.* Accordingly, in the fundamental equation of soil-forming factors

$$s = f(cl, o, r, p, t, \cdots) \tag{4}$$

the letter o refers to the biotic factor rather than the number, proportion, and yield of the species actually growing on the soil. The latter results from the interaction of the biotic factor and the other soil formers.

The problem that now lies before us centers around the following question: Knowing the biological complex existing on a given soil, how can we evaluate the corresponding biotic factor? Or, to express the same question in terms of our previously mentioned experiment: Knowing the dependent variables b_1, b_2, b_3, etc., how can we estimate the independent variable or biotic factor B? It will prove profitable to conduct this inquiry separately for each pedologically important group of organisms, namely, microbes (o_m), vegetation (o_v), animals (o_a), and man (o_h).

Microorganisms.—Relatively little systematic knowledge is at hand regarding the distribution of microorganisms in various soil types, but what little we know indicates definitely that each soil possesses its own characteristic microbial population. Furthermore, it has been established by countless experiments that changes in the properties of a soil are accompanied by changes in its microbiological constitution. Strongly acid soils, for instance, are void of the nitrogen-fixing bacteria *Azotobacter*. However, when neutralized, these soils, as a consequence of natural reinoculation, soon abound in this important group of microorganisms. In nature, the facilities for a wide redistribution of microbes are practically unlimited, largely because of the small size and ease of transportation of the organisms. Air movements, precipitation, and dust storms bring about a continuous reinoculation of soils. We are, probably, not far from the truth by assuming that nearly all soils at some time or another receive samples of most of the soil microorganisms. This belief leads to the assumption that in the vast majority of soils the same microbiotic factor is operative. Hence, within a large region, the microbiological component of the biotic factor o in Eq. (4) may be taken as constant, being nearly identical for all soils. The approximate composition of the microbiotic factor simply corresponds to the sum of the species of the microorganisms found within the region. According to this viewpoint, the individual microbial populations of each soil within the region are merely the consequence of the great variety of constellations of the remaining soil-forming factors.

To forestall any misunderstandings, we should like to emphasize that the discard of each specific microbial complex as a separate soil-forming factor does not in any way minimize their practical and scientific significance in the intricate mechanism of physical, chemical, and biological processes. However, this book is a treatise on soil-forming factors and does not undertake to study soil-forming processes.

Vegetation.—In pedological research we may wish to evaluate the specific effect of various plant species or types of vegetation on soil formation, or we may desire to study soil development under conditions of constancy of the vegetational factor. Both problems merit separate consideration.

a. If we wish to contrast the influence of two plant species or plant associations on soil formation, all other soil formers must be kept constant. This postulate follows directly from the specific equation

$$s = f\ (\text{vegetation})_{cl,r,p,t,\ldots}$$

In the field it must be shown that climate, topography, parent material, and time are similar for the two types of vegetation to be compared; in other words, the two species or groups of species must appear in the role of independent variables or biotic factors. As an example, one might point to the multitude of agricultural crops that are frequently planted side by side under identical conditions of climate, topography, parent material, and time. Each crop or sequence of crops (rotation) acts as a biotic factor.

b. The problem of treating the biotic factor as a "constant" occurs whenever the general equation

$$s = f(cl, o, r, p, t, \cdots) \tag{4}$$

is to be solved for either cl, r, p, or t. Expressing s as a function of the climatic variable, we have, in the preceding chapter, written Eq. (4) as follows:

$$s = f\ (\text{climate})_{o,r,p,t}$$

Here we are confronted with the task of evaluating o as a constant while climate varies. The general solution of this problem has been suggested by Overstreet.*

FIG. 100.—Illustration of a simple biotic factor. Distribution of *Atriplex* in a desert landscape (playa).

The reader's attention is directed to Fig. 100, which shows a section of a desert landscape. The white barren area represents a playa, *i.e.*, an old lake bed incrusted with alkali salts. The borders and sand elevations are covered with saltbush (*Atriplex*). Although the playa lacks vegetation, it cannot be said that it lacks a biotic factor, for the seeds of the saltbush certainly are scattered over the salt beds. Plant growth on the playa is nil because of unfavorable physiologic conditions and not because of absence of potential vegetation. Both the playa and the elevations possess the same biotic factor, namely, saltbush. It is conceivable that additional plant species reach this particular landscape but fail to survive because of lack of sufficient water.

* Personal communication.

These species also should be included in the description of the biotic factor.

Figure 101 portrays a segment of the semiarid portion of the Coast Range in California. Oaks grow in the canyons and depressions, and grasses cover the slopes and ridges. Assuming constancy of climate and parent material, the distribution of grasses and trees is clearly a function of topography and therefore constitutes a dependent variable. In this specific landscape the biotic factor consists of oaks *and* grasses.

FIG. 101.—Vegetational factor consisting of oaks and grasses (Coast Range, California).

In certain parts of the limestone regions of the Alps, slightly weathered rock has a cover of basophilous *Dryas octopetala*, whereas adjacent acid soil derived from the same material as a result of intensive leaching supports acidophilous *Carex curvula*. Since the seeds of both species have access to both kinds of soil, the biotic factor is the same for both soils, namely, *Dryas* plus *Carex*. In this instance, the differentiation in the actual vegetational cover is brought about by the factor time or the degree of maturity. Ultimately, the slightly weathered material will be converted into a strongly acid soil with a simultaneous displacement of *Dryas octopetala* by *Carex curvula* (compare page 216).

Proceeding to more general cases, we may enunciate the following ecologic principle: If in any region the plant species or plant communities are dependent variables, their distribution being a function of one or of several soil-forming factors, the biotic factor may be obtained, as a first approximation, by enumerating all plant species growing within the area.

A more detailed discussion of the dependent and independent nature of vegetation will be presented in succeeding sections.

Animals.—A similar line of thought may be applied to the influence of animals in soil formation. Because of lack of sufficient observational data covering wide areas, the discussion of animal life is omitted in this treatise.

Man.—Like vegetation, man may appear in the role of a dependent as well as an independent variable. With respect to man's dependency on soil-forming factors, some scientists (43) go so far as to attribute the origin of different human races to the influence of soil and climate. Hilgard and especially Ramann (47) have emphasized the relationships between social structures and soils, as indicated by despotic governments found in ancient irrigation areas and the more democratic institutions developed on soils of humid regions. Huntington stresses the importance of weather on health and efficiency of human beings and traces the rise and fall of certain civilizations to climatic changes. Abbott (1) calls attention to widespread nutritional anemia among children who live on home-grown food from poor soil that is deficient in iron. Medical science in general knows countless examples of correlations between pathology of man and environment.

A considerable number of human influences on soil appear to stand in no direct relationship to soil-forming factors. Lands in all parts of the world are plowed and are subjected to numerous cultural treatments. Stable manure is added to the soil wherever cattle are raised. Crops are harvested universally. Deforestation occurs on all continents, and burning is practiced whenever needs arise. In all these enterprises, man acts, as far as the soil is concerned, as an independent variable or soil-forming factor. Some of the consequences will be discussed in subsequent sections.

B. VEGETATION

1. Vegetation as a Dependent Variable

In order to clarify the dual role of vegetation as a dependent and an independent variable, we first present a series of relationships that emphasizes the combined effect of soil-forming factors on the nature of the plant cover. To prepare the way for subsequent discussions, it may prove helpful to become familiar with certain botanical terminologies for vegetational complexes.

Plant Associations and Plant Climaxes.—The concept of a plant association, or plant community, has become important in the study of soil-plant relationships. It has been discovered that the development of these units of vegetation is closely related to soil-forming processes.

Flahault and Schröter define an *association* as a plant community of definite floristic composition. Braun-Blanquet (9)

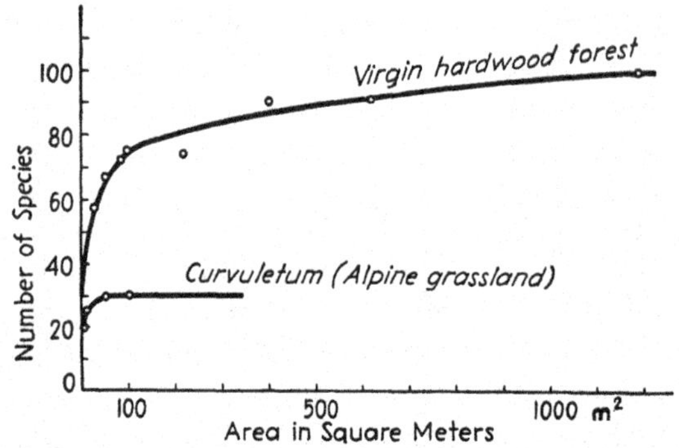

Fig. 102.—Relationship between area and number of species of two plant associations.

writes, "Pieces [areas] of vegetation with similar combinations of species are united into abstract types. These types are the *associations*, the separate pieces being called the *individuals*, or examples of the association." Other investigators, *e.g.*, Weaver and Clements (65) and Raunkiaer (48), give somewhat different definitions.

The *number of species in a plant association* varies with the area but tends to reach a limiting value. The alpine grass association *Curvuletum* contains about 30 different species that are spatially close together. A quadrate of 1 sq. m. selected at random within the area of the association will contain most of the species. On the other hand, a virgin hardwood forest, in Posey County, Indiana (13), embraces over 100 species, and the size of the *minimalraum* (minimum area that contains all the species) is much greater, as shown in Fig. 102. The areal extent of major plant associations may comprise thousands of square miles.

Land made barren by cultivation, erosion, landslides, retreating glaciers, or volcanic eruptions offers exceptional opportunities for the study of the *development of plant associations* and their relations to soil. In the initial phase of the development of the vegetation, when only a fraction of the soil is covered by plants, competition is not keen. The plant cover is haphazard in composition, and the associations are of a transitory nature. As the covering becomes denser, the struggle for life takes on a more competitive form. Some species succumb, some stay, others become dominant. After a number of years, the vegetational cover is stabilized and consists of those plant immigrants that are best adapted to the environment. In arid regions, neutral or alkaline soil reaction is likely to prevail, and acidophilous plants will be crowded out. In districts where salt accumulations occur, the halophytes only will survive. In humid regions, acid soils evolve, and the basophilous species are ultimately suppressed by acid-tolerant plants with low soil-fertility requirements. These final plant communities are called "climax associations," or "*climaxes*" (14). The soil beneath the climax vegetation often has the profile characteristics of mature soil types.

TABLE 51.—AREAS OF NATURAL VEGETATION IN THE UNITED STATES (57)
PERCENTAGE FIGURES
(Total area = 1,903,000,000 acres)

Vegetation		Group total	
Northern desert shrub	= 10	Total desert shrub	= 14
Southern desert shrub	= 4		
Tall grass (prairie)	= 16	Total grassland	= 38
Short grass (plains)	= 14		
Other grasses	= 8		
Coniferous forest	= 22	Total forest	= 48
Hardwood forest	= 21		
Woodland	= 5		

Broad Groups of Vegetation.—From the viewpoint of soil formation, the most important classes of vegetation are the forests, the grasses, and the desert shrubs. The forests predominate in humid regions, the desert shrubs in arid regions, and virgin grasslands occupy the transition zone. The areas of these major divisions and some of their subgroups in the United States are given in Table 51. The areas refer to the natural vegetation, *i.e.*, the type of land cover that supposedly existed before the white settlers arrived. Nearly all the Eastern grasslands have

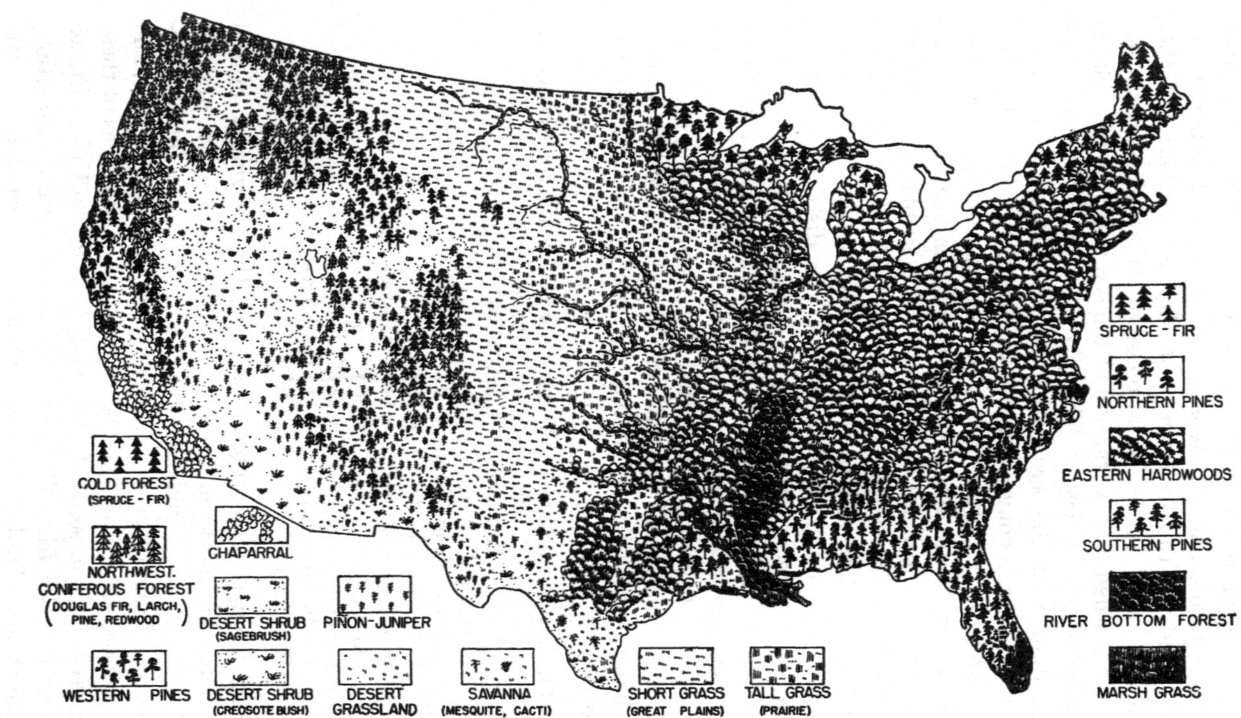

FIG. 103.—Natural vegetation map of the United States. (*Adapted from C. Sauer, Man in Nature, Charles Scribner's Sons, and Atlas of American Agriculture, Part I, Section E, Natural Vegetation.*)

TABLE 52.—ORGANIC-MATTER PRODUCTION BY VEGETATION
(Roots not included unless stated)

Type of vegetation	Annual production		Author	Remarks
	Tons per acre	Metric tons per hectare		
Alpine meadows	<0.22–0.40	<0.5–0.9	Swederski (60)	Clipped quadrats, Poland; air dry (includes mosses and lichens)
Short-grass prairie	0.71	1.6	Clements and Weaver (16)	Clipped quadrats, Colorado; air dry
Mixed tall-grass prairie	2.22	5.0	Clements and Weaver (16)	Clipped quadrats, Nebraska; air dry
Tall-grass prairies	0.85–1.73	1.9–3.9	U. S. Census	Annual prairie hay values, Middle West
Average forest (leaves, wood, twigs)	2.67	6.0	Henry (67)	German forests, air dry
Beech: Wood	1.42	3.2	Ebermayer (67)	Central Europe, air dry
Beech: Leaves	1.47	3.3		
Pine: Wood	1.42	3.2	Ebermayer (67)	Central Europe, air dry
Pine: Needles	1.42	3.2		
White-pine needles	2.09	4.7	Lunt (40)	Connecticut, 1934, oven dry
Beech-birch-aspen forest	2.85	6.4	Henry (67)	Leaf litter only; Germany, air dry
Tropical primeval forest (leaves, trunks, roots)	11.1	25	Hardon (21)	Java, estimate, moist
Tropical legumes	24.4	55	Koch-Weber (21)	Dutch East Indies, moist
Tropical savannas	13.3	30	Vageler (63)	Estimate, moist?
Monsoon forest	22.2	50	Vageler (63)	Estimate, moist?
Tropical rain forest	45–90	100–200	Vageler (63)	Estimate, moist?

been plowed. Of the original timber, only about 260,000 sq. miles of merchantable forest land, mostly in the West, are left.

Production of Organic Matter.—In the preparation of Table 52, an attempt has been made to collect numerical data regarding the yields of natural vegetation. Unfortunately, the data are so widely scattered throughout the literature of the world, and the methods of measurements are so variable that the table is only of fragmentary nature. In spite of its incompleteness, certain significant conclusions appear possible. Within the temperate region, a good stand of forest produces more organic matter than a good virgin prairie, when the leaf production and the wood growth are added together. The leaf production alone is approximately of the same order of magnitude as the hay production of tall-grass prairie. A further point of interest is the enormous growth of vegetation in the tropics. In contrast, the virgin grasslands of high mountains yield low production figures.

A further shortcoming of Table 52 is due to the fact that the data refer only to organic-matter production aboveground, yet the nature and extent of the root systems are equally important, especially from the viewpoint of soil formation. Recently, Kramer and Weaver (35) have published data on root production in the surface four inches of soil underlying various plant species (Table 53). One of the surprising results is the high value for certain grasses. They produce much more organic matter underground than aboveground.

TABLE 53.—DRY WEIGHTS OF LIVING UNDERGROUND PLANT PARTS IN THE SURFACE 4 INCHES OF SOIL (*Kramer and Weaver*)

Plant	Average weight, grams per 0.5 sq. m.	Weight at maturity, grams per 0.5 sq. m.	Pounds per acre
Slough grass	624	742	13,240
Buckbrush	645	680	12,134
Big bluestem	462	462	8,200
Little bluestem	376	376	6,600
Bluegrass	...	269	4,800
Brome grass	185	220	3,926
Alfalfa, old	158	196	3,497
Wheat (lowland)	51	75	1,338
Corn (maize)	31	65	1,160
Peas	...	14	250

During the fall of 1932, Weaver, Hougen, and Weldon (66) selected representative square meters of typical virgin prairies in Nebraska. The soil was removed to the depth of penetration of the grasses for the purpose of determining the *relation of root distribution to the amount of organic matter in prairie soil.* The roots were separated from the soil, and the organic matter in both roots and soil was determined. Measurements of the volume weight of the soil permitted accurate weight calculations (Table 54). Except for the surface level of soil, there is an approximate linear relation between the amount of root material and the amount of organic matter in the various soil horizons. Roots and rhizomes constitute about one-tenth of the total organic matter in the surface six inches of soil; in the deeper sections, the proportion decreases gradually from 3 to 4 per cent in the second six inches to 1 per cent in the fourth foot. The organic-matter content and the total nitrogen content of the plant roots vary from 83.1 to 91.7 per cent and from 0.66 to 0.83 per cent, respectively.

TABLE 54.—ORGANIC MATTER AND NITROGEN RELATIONS IN ROOTS AND SOIL (*Weaver, et al.*)
(One square meter area; Lancaster loam)

Depth	Grams organic matter*		Grams total nitrogen	
	In soil†	In roots	In soil†	In roots
0– 6 in.	7,670‡	657	382‡	6.13
6–12 in.	6,480	182	334	1.39
1– 2 ft.	6,840	160	400	1.26
2– 3 ft.	2,030	74.0	203	0.62
3– 4 ft.	1,000	11.0	120	0.10

* Determined by a modification of the hydrogen peroxide method of Robinson.
† Includes the roots.
‡ This includes also the rhizomes of the grasses, all of which occurred in the surface six inches of soil.

Regarding forests, German investigators report that the weight of the underground mass of vegetation is from 20 to 30 per cent of the total mass [Weber (67)]. According to Zederbauer (69) the ratio of the weight of roots to that of the aboveground parts in the case of fourteen-year-old spruce is about one to four. Scholz (55) dug 137 profile holes in the Black Rock Forest, an abused forest 50 miles north of New York City, and found the greatest number of tree roots in the upper eleven inches of the profiles. The number decreased rapidly in the 11- to 18-in. zone, and only an occasional root was found below 18 in.

Vegetation as Controlled by Climate.—A vast amount of literature exists on the relationship between vegetation and climate, particularly with regard to the distribution of species. It will be recalled that Transeau's work in 1905 led to the concept of the precipitation-evaporation ratio. In fact, much of the impetus in certain phases of climatic research originated with botanists. No student of ecology can afford to overlook the voluminous work of Livingston and Shreve on the distribution of

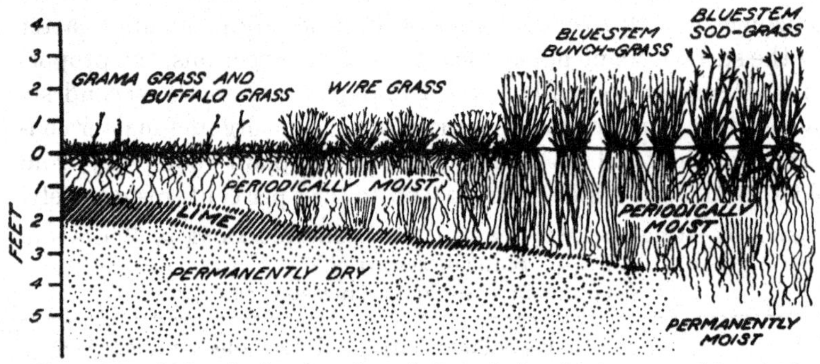

Fig. 104.—A sketch showing the relation of the plant communities to the depth of penetration of soil moisture in the Great Plains area. [*From Shantz* (56).]

vegetation in the United States of America as related to climatic conditions. The complexity of plant-climate relationships is clearly brought out in this work. Unlike soil types, plant species are very sensitive to extremes of weather, and for this reason mean annual climatological values are, generally speaking, of limited significance.

The drawing by Shantz (56) that is reproduced in Fig. 104 is instructive from a *qualitative* point of view. It shows a transect from the Rocky Mountains across the central Great Plains to the Missouri River. The succession of plant communities in relation to increasing rainfall (left to right) is clearly brought out. In the western part (left), the grasses are but a few inches in height; whereas, in the eastern part, the bluestem often reaches a height of over 3 ft. The root development varies accordingly. Shantz writes:

The plant distribution is correlated with the depth below the surface of the layer of carbonate accumulation. Where this depth is less than 2 ft., the plains type of vegetation predominates. Where greater than about 30 in. or where lacking entirely, the prairie type of grassland occurs. The important point here is the depth of soil periodically moistened by rainfall and the total moisture supply available. Short grass characterizes areas where each season all available soil moisture

TABLE 55.—MOISTURE BOUNDARIES OF PLANT ASSOCIATIONS SHOWN IN FIG. 104

Region	Vegetation	Approximate annual rainfall, inches
Eastern Colorado.....	Grama-buffalo grass	Below 17
Western Kansas......	Wire grass	17–21
Central Kansas.......	Bluestem bunch grass	21–27
Eastern Kansas and western Missouri...	Bluestem sod grass	27–38

is consumed by plant growth. All available soil moisture is also consumed along the western edge of the tall grass. Over the tall-grass area as a whole, however, moisture during the rainy period penetrates so deep into the soil that it is not all recovered and brought to the surface by plants. Consequently, the carbonates are carried down and away entirely with the drainage water. At the beginning of the growth period, the soil is moist to the layer of carbonate accumulation, the equivalent of from 4 to 6 in. of rainfall in the tall grass, and from 2 to 4 in. in the short grass.

On the basis of the vegetational map of Shantz and the official rainfall maps, the climatic limits of the plant communities have been tabulated in Table 55. The values refer to northern Kansas, approximately along the 52°F. (11°C.) annual isotherm.

Quantitative information on vegetative growth and climate is very limited. Of the few reliable data at hand, the measurements of Clements and Weaver (16) given in Table 56 deserve special consideration. These two ecologists carefully clipped the vegetation from selected quadrates and weighed the dry matter. The data corroborate the general qualitative observation that plant production increases from west to east in proportion to higher rainfall.

If one traverses the Great Plains area from east to west, *i.e.*, from regions of high to those of low rainfall, one is impressed by the change in the appearance of the *corn fields*. In the East, the stalks are large; in the West, they are small. This is caused

TABLE 56.—RAINFALL AND PLANT PRODUCTION IN THE GREAT PLAINS AREA. VIRGIN GRASSLANDS (*Clements and Weaver*)

Locality	Average annual rainfall, inches	Grams of dry matter per square meter		
		1920	1921	1922
Burlington, Colo.	17.4	183	353	224
Phillipsburg, Kans.	22.7	378	402	311
Lincoln, Neb.	28.5	458	603	447

in part by differences in precipitation and in part by differences in soil fertility. Table 57 published by Salmon (52) proves that in Kansas annual rainfall may be closely correlated with corn yield.

TABLE 57.—EFFECT OF ANNUAL PRECIPITATION ON THE YIELDS OF LARGE, MEDIUM, AND SMALL VARIETIES OF CORN (ZEA MAYS) (*Salmon*)

Variety	Number of days to mature	Average yield, bushels per acre		
		Eastern Kansas 27–35 in. of rainfall	Central Kansas 21–27 in. of rainfall	Western Kansas 16–21 in. of rainfall
Kansas sunflower	125	44.3	20.1	10.1
Pride of Saline	115	42.1	21.8	13.3
Freed white dent	105	38.8	27.2	18.3

No data could be found in the literature that expressed quantitatively the effect of *temperature* on organic-matter production under natural conditions. Of course, it is generally known that in cold regions plant life is hampered, whereas in the tropics it may assume fantastic dimensions.

In order to arrive at some quantitative index, the *yields of wild grasses* were plotted against temperature (Fig. 105). The data were taken from the agricultural statistics provided by United

States Census. Every dot in the graph represents the mean for the average yields of a county for the years 1910, 1920, and 1930. Only tall-grass prairies situated east of the 97th meridian were included. There is surprisingly little variation in organic-matter production from Canada to the Gulf of Mexico; most of the values are scattered through the interval 0.8 to 1.6 tons per acre (1.8 to 3.6 m.-tons per ha.). The yields tend to be highest in the corn belt, a feature that probably is due to superior soils. There is no indication in these data that plant production increases expo-

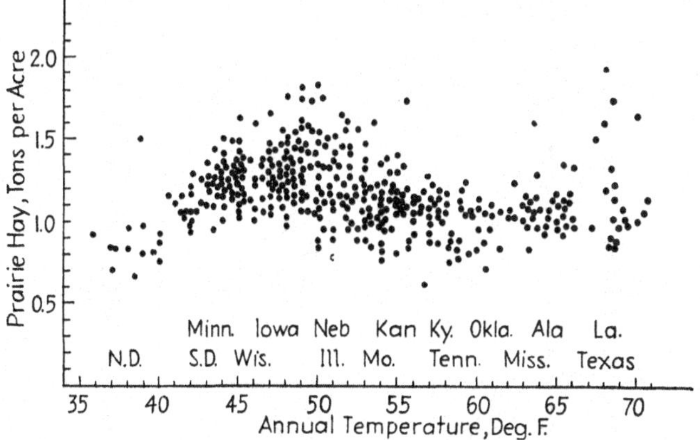

FIG. 105.—Average annual yields of prairie hay in relation to mean annual temperature.

nentially from north to south as stated by a number of scientists who base their claims on van't Hoff's law. Apparently adverse conditions of soil and perhaps moisture operate against the basic temperature principle.

Combined Effects of Climate and Soil on Corn Yields (29).—The curve of corn (*Zea mays*) yields from north to south assumes an instructive shape. Within a belt of relatively uniform precipitation-evaporation ratios that extends from eastern North Dakota through Minnesota, Iowa, Missouri, and Arkansas to Louisiana, average corn yields exhibit a rising trend in the North and, after reaching a maximum in the corn belt, decline sharply toward the Gulf Coast (Fig. 106). The graph also contains the nitrogen-temperature curves of upland, terrace, and bottom-land soils of the same region. The parallelism between average corn yield and average soil nitrogen in the states south of Iowa is the most prominent feature of the graph. This proportionality may be expressed by the equation

$$Y = 224.6N - 3.31 \tag{35}$$

Y denotes the average corn yield (10-year period) in bushels per acre, and N indicates the average total nitrogen content of surface soils. The correlation is nearly perfect, having a coefficient of $r = +0.995$.

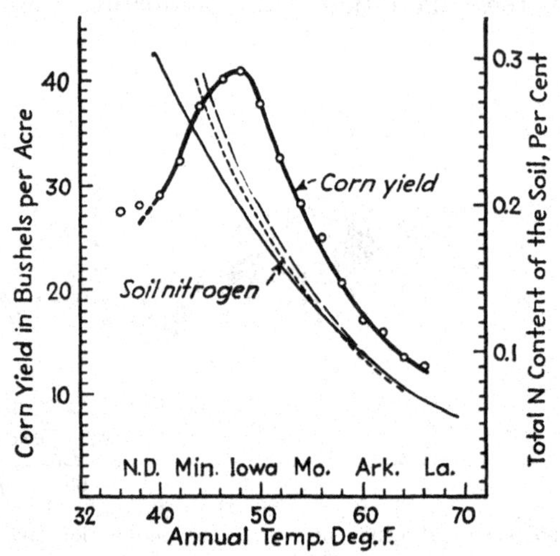

Fig. 106.—Showing the relation of corn yield and nitrogen content of upland terrace, and bottom-land soils to the annual temperature.

The most probable explanation of the dependency of corn yields on latitude appears to be as follows. In the North, low temperatures limit the development of the corn plant. The ascending branch of the curve reflects the beneficial effect of generally higher temperatures on the yields of corn: that is to say, we observe the results of a *direct* influence of climate on corn production. On purely climatological grounds, no good physiological reason is apparent why the yields should decline sharply in the region south of Iowa. Here, the well-established fact that corn is a heavy nitrogen feeder and demands a high level of fertility in general must be taken into account. In other words, soil takes the leading part in determining the yield of corn. South of Iowa we are confronted with an *indirect* effect of climate on corn production. High temperatures lower the nitrogen content of the soil and, coupled with a low amount of available nutrients, limit the quantity of plant growth.

The contention that south of the corn belt soil rather than climatic conditions controls the average yield of corn receives confirmation from several independent sources. In the Missouri and Illinois sections of the corn belt the total soil nitrogen represents a convenient numerical index of general soil fertility, and the yields of corn, within comparable climatic conditions, vary accordingly. In the South, Funchess (17) has shown conclusively that corn yields may be increased appreciably by green manuring with legumes or by application of nitrogenous fertilizer. Moreover, Funchess maintains that fertilization without nitrogen will not result in large crop yields.

The corn yield-latitude curve provides a good illustration of the direct and indirect effect of climate on the yield of a single plant species.

Parallelism between Plant Successions and Soil Development.—As soon as fresh rock is exposed to a biological environment (biotic factor), certain organisms, notably bacteria, take possession of it. Blöchliger (6) found that the number of bacteria on barren limestone rocks in the Alps is greater the more advanced the degree of rock decomposition (Table 58). Bacteria were abundant even on the surface of the unweathered rock. These organisms were saprophytes, which utilize the carbon and nitrogen content of dead organic substances. Anaerobic bacteria

TABLE 58.—NUMBER OF BACTERIA PER ONE GRAM LIMESTONE ROCK MATERIAL (BARREN SURFACE)

Type of material	Bacteria grown on gelatin	Bacteria grown on agar
Surface of rock and minute crevices	73,320	67,790
Scratched-off parts	452,000	267,800
Rock debris	762,778	813,333
Soil material	2,306,250	2,723,750

were found even in the smallest cracks. No nitrogen-fixing bacteria of the type such as *Azotobacter chroococcum* could be isolated, but anaerobic nitrogen-fixing species of the group *Bacillus amylobacter* were observed. In the more advanced phases of rock decay, nitrifying bacteria were numerous, and fungi and *Actinomyces* were quite abundant. Evidently microorganisms play a primary role in biological rock decay.

According to Braun-Blanquet and Jenny (28), on physically disintegrated limestone rock of the central Alps, *Dryas octopetala* and *Carex firma* are the pioneers of vegetation. Their roots hold the soil particles together, check erosion, and thus permit the beginning of soil-profile formation. Carbon dioxide from root excretion and from microbiological decomposition of plant material increases the solubility of the carbonates. Weathering

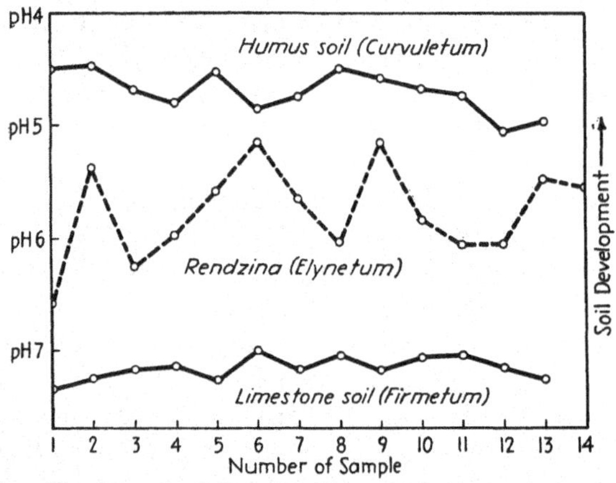

FIG. 107.—pH values of soil transects through plant associations representing initial (*Firmetum*), intermediate (*Elynetum*) and climax (*Curvuletum*) stages of plant successions.

is accelerated. The insoluble aluminosilicates contained in the original limestone rock accumulate, forming silt and clay; phosphorus and potassium increase relatively because they are leached less than the calcium amd magnesium bicarbonates. The pH of the soil fluctuates between 7.0 and 7.4. The pH values of soil transects through typical pioneer plant associations give almost straight lines parallel to the X-axis (Fig. 107). This state may be termed the initial phase of vegetation and soil development.

As weathering and soil development continue, more plants with higher fertility requirements intrude upon the pioneer associations. *Elyna myosuroides*, *Festuca violaceae*, *Sesleria coerulea*, and *Trifolium thalii*, all of which are characterized by abundant vegetative growth, begin to dominate and to form new associations. In turn, these plants enhance soil development. The increased production of organic matter results in the augmentation of the humus content of the soil; the profile becomes dark and deep and is now classed as rendzina. The soil reaction varies from neutral to slightly acid and fluctuates considerably, as is evident from the transect line in Fig. 107. Both vegetation and soil are in a state of great transformation that represents an unfinished, intermediate phase of plant and soil development. From an agricultural viewpoint, soil and vegetation are in optimum conditions, and represent the famous fertile alpine meadows (Table 59).

At such high altitudes, the prevailing low temperature facilitates the further accumulation of humus. The high precipitation results in a pronounced leaching of bases, and hydrogen ions become the dominating soil cations. Basophilous and nitrophilous plants disappear and acid-tolerant species, such as *Carex curvula*, increase in number. The intruders are small in size, produce less organic matter, and the new associations contain a smaller variety of species. From an agricultural point of view, both soil and vegetation degenerate, although, genetically speaking, their development represents a more stable state. The vegetational climax of the region is characterized by the *Curvuletum* with *Carex curvula* dominating. The corresponding soil climax is known as "rendzina podsol" or the "alpine humus soil," depending on altitude. These soils are strongly acid (pH 4.5 to 5) and more homogeneous than those of the preceding rendzina stage. Unless there occurs a major change in one of the soil-forming factors, soil and plant climaxes remain relatively stationary.

2. Vegetation as an Independent Variable

If we wish to study soil formation as a function of vegetation, we write the general equation of soil-forming factors in the following form:

$$s = f\,(\text{vegetation})_{cl,p,r,t,\ldots} \qquad (36)$$

All vegetation growing on a soil affects the soil but is not necessarily a soil-forming factor. Only that aspect of vegetation is considered a soil-forming factor that cannot be correlated with

TABLE 59.—PLANT SUCCESSION AND SOIL DEVELOPMENT IN THE ALPS (*Braun-Blanquet and Jenny*)

Development of vegetation			Development of soils				
Stage	Plant associations	Relative production of organic matter, per cent	pH	Total acidity, milliequivalents per 100 g.	Organic matter of surface soil, per cent	Magnitude of humus layer, centimeters	General description of soils
Initial	*Firmetum*	40	7.2–0.04	0	0–30	0–5	Disintegrated and slightly weathered limestone rock
Transitional	*Elynetum, Festucetum*	100	6.1–1.07	1.3	20–50	10–30	Rendzina profile, leaching of lime, humus accumulations
Climax	*Curvuletum*	60	4.8–0.03	7.2	20–60	10–100	Alpine humus soil, leached profile, acid humus

climate, parent material, topography, and time. This aspect (the biotic factor *o* in the general equation) becomes particularly simple and easily evaluated wherever man controls the vegetational cover, as in all agricultural and many silvicultural practices. Under natural conditions, it is often exceedingly difficult to estimate reliably the vegetational component of the biotic factor. A classic example of this problem is provided by the prairie-timber transition zone in which the two great divisions of vegetation, virgin prairie and virgin forest, live adjacently in areas of apparently identical climate, parent material, topography, and age.

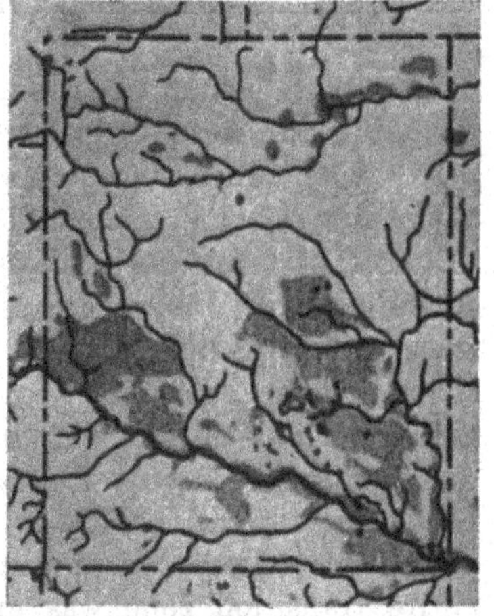

FIG. 108.—Original distribution of forest (dark areas) and prairie (light areas) in Tama County, Iowa. Note that both prairie and forest occur on the uplands as well as in the drainage channels. (*Iowa State Planning Board.*)

The Prairie-timber Transition Zone.—The underlying causes of the coexistence of prairie and timber have been a puzzle to many a scientific mind. A recapitulation of the salient facts and a brief enumeration of explanations that have been suggested are here inserted in order to foster interest among pedologists.

The prairie-timber transition belt (18) has been described in European Russia, Siberia, Canada and the Central part of the United States. In the last country, it occupies a strip several hundred miles in length. The belt is composed of vast areas of prairie framed with extensive tracts of timber. Over the entire region, rainfall is high enough to permit luxuriant tree growth.

The explanations of the coexistence of prairie and forest may be arranged into two groups. The first comprises those explanations that are based on the assumption of a single biotic factor (prairie plus forest) for the entire area. In consequence, the distribution pattern of the two vegetational complexes is considered a dependent variable, the character of which is determined by the soil and its environment. Level topography, poor water permeability and a high-water table are said to be favored by prairies, whereas rolling lands with good drainage are preferred by timber. However, numerous exceptions are known. Huntington (27) believes that the present-day climatic differences suffice to explain the vegetational pattern. He compares rainfall and temperature of Des Moines, Iowa, a prairie district, with those of Akron, Ohio, which lies in the timber phase. During certain months, Akron has from 1 to 2 in. more rainfall than Des Moines; hence, Huntington concludes, the slightly more continental nature of the Iowa climate accounts for the absence of forest.

The second group of explanations rests on the belief that two distinct biotic factors are operative or have been operative in the past. The adherents of this school of thought call attention to the role of higher animals and man as preventers of the establishment of forest. Pasturing buffaloes or repeated grass fires spreading from Indian camps and villages have destroyed the tree seedlings as rapidly as they were produced. These activities of the higher organisms would presumably result in the maintenance of two distinct vegetational factors, forest and prairie.

An entirely different theory has been advocated by Gleason (19) in his Vegetational history of the Middle West. The prairies are considered relics of prehistoric dry periods during which the entire transition zone was grassland. A climatic change has since taken place. The climate is now more humid and favorable for forests that encroach upon erosion channels. Vegetational equilibrium has not yet been reached. Ultimately all prairie will succumb.

Although the explanation of the prairie-timber transition zone presents an especially difficult problem, for the cases that we are about to discuss all observations seem to justify the assumption of two distinctly different and independent biotic factors. In all subsequent discussions, the two biological complexes, prairie and forest, will be treated as different soil-forming factors.

Prairie and Timber Profiles in Minnesota.—Rost (50, 51) has analyzed virgin timber and virgin prairie soils in Rice County in southern Minnesota. The average annual temperature is 45°F.; the annual *NS* quotient value is approximately 330. The soils of the Carrington series have developed under identical conditions of climate, parent material (late Wisconsin moraine), topography, and, presumably, time. One set of profiles, the

Carrington silt loam, has developed under bluestem sod, the other, the Carrington loam, under oak-hickory hardwoods.

A complete listing of Rost's numerous analyses would consume too much space. Some of the most significant results are condensed in Table 60 by making use of molecular ratios.

The uniformity of parent material is indicated by the amount of coarse gravel, which is practically identical for both prairie and timber. The moisture equivalent, which is a function of the water-sorption power of clay and humus, is higher in the prairie

TABLE 60.—COMPARISON OF PRAIRIE AND FOREST PROFILES UNDER CONDITIONS OF CONSTANT ENVIRONMENT (*Rost*)

Constituents	Vegetation	Depth			
		1–6 in.	7–12 in.	13–24 in.	25–36 in.
Coarse gravel, per cent	Prairie	0.88	1.04	1.87	3.03
	Forest	0.52	0.91	1.64	2.15
Moisture equivalent	Prairie	29.6	27.7	25.7	25.9
	Forest	24.4	20.6	21.9	20.5
Organic carbon, per cent	Prairie	4.48	3.19	1.78	0.77
	Forest	3.06	1.46	0.81	0.50
Carbon-nitrogen ratio, C:N	Prairie	11.7	10.7	10.7	13.3
	Forest	12.6	12.0	12.4	12.8
$\frac{SiO_2}{Al_2O_3}$	Prairie	11.9	11.5	11.3	10.3
	Forest	13.6	12.3	11.0	11.1
$\frac{CaO}{Al_2O_3}$	Prairie	0.22	0.22	0.20	0.31
	Forest	0.21	0.17	0.15	0.17
$\frac{10\ CO_2}{Al_2O_3}$	Prairie	0.16	0.11	0.29	2.02
	Forest	0.24	0.13	0.10	0.08

profile, but in no instance is there any sign of marked translocation of clay, which would indicate a typical *B* horizon. In general, the chemical alterations in the two profiles are slight. It appears that differentiation into horizons has not been very intense, in spite of the considerable lapse of time (from 10,000 to 20,000 years) since the retreat of the glaciers.

Regardless of the limited degree of weathering and soil development, the role played by vegetation is unmistakably preserved. Total organic carbon and, accordingly, organic matter is more abundant in the prairie than in the forest soil. The same holds true for total nitrogen. The carbon-nitrogen ratio is wider for the forest, a feature that seems to be generally characteristic of forest soils (25). The rather abrupt change of organic matter with depth is also more or less consistently associated with tree vegetation.

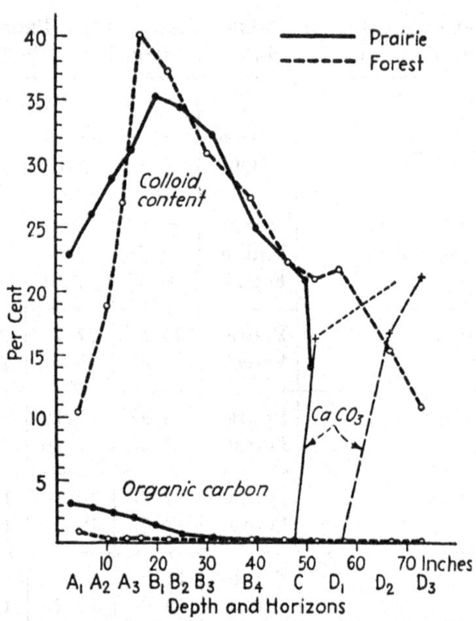

FIG. 109.—Comparison of soils developed under prairie and under forest (Illinois). All other soil-forming factors are identical.

The silica-alumina ratio is slightly higher under forest than under prairie, indicating that translocation of alumina has been hastened. The behavior of CaO reveals a similar trend of leaching. The CaO/Al_2O_3 values are high for prairie and low for timber, especially in the lower horizons. The values for carbon dioxide, although low in both cases, corroborate these tendencies.

Summarizing the influence of the plant cover on profile differentiation, the following two points stand out. In the first place, the distribution of organic matter is conspicuously different for prairie and for timber. The latter has a large proportion of its carbon content in the surface layer. In the second place, the translocation of mineral substances is greater under timber than under prairie. From the data at hand, one would conclude that under equal climatic circumstances a deciduous forest cover stimulates leaching and accelerates soil development.

Soil Studies in Illinois (10).—In the west central part of the state of Illinois, extensive till plains were formed by the retreating glaciers of the Illinoian Age. They are now covered by loess

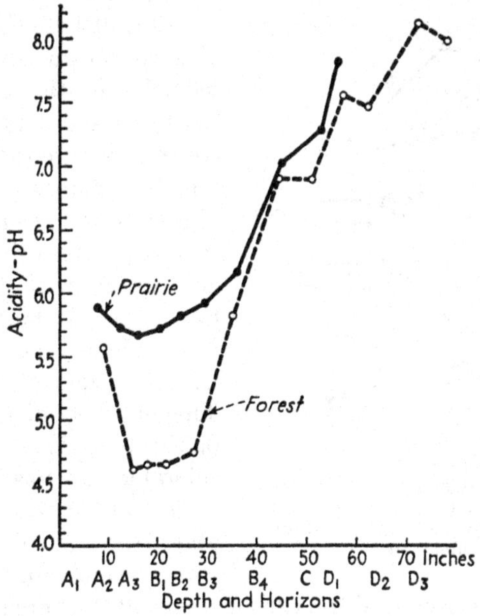

Fig. 110.—Comparison of soils developed under prairie and under forest (Illinois). All other soil-forming factors are identical.

deposits of over 10 ft. in thickness. The loess probably originated from the Iowan glaciation and bears a luxuriant prairie and timber vegetation. Comparison of the Grundy silt loam (prairie) and the adjacent Rushville silt loam (forest) permits quantitative isolation of the soil-forming factor—vegetation.

The *climate* is semihumid with a rainfall of 30 to 35 in. and with an annual temperature of 53°F. The value of *NS* quotient is approximately 350. Owing to extremely level topography, the surface drainage is poor, but the underdrainage is satisfactory because of the open structure of the loess. Analyses of the two soil types mentioned are graphically assembled in Figs. 109 to 111. They refer to two profiles taken from cultivated fields.

Organic Matter.—In harmony with previous statements, the prairie soil is much richer in humus than the forest soil. The difference amounts to several hundred per cent.

Inorganic Colloid Content.—To a depth of 50 in., the average content of colloidal clay particles ($<1\mu$) is nearly the same in the two profiles, 28 per cent for prairie and 26 per cent for timber. Significant differences are found in the distribution patterns. The concentration of the clay particles in the B_1 horizon is most marked under the forest cover, a feature that is probably due to enhanced translocation of clay from the surface horizon to lower parts of the profile. In accordance with this explanation, the *A* horizons of the Rushville are lower in clay than the *A* layers of the Grundy. The relatively high clay content of the *C* horizon suggests considerable clay formation *in situ*.

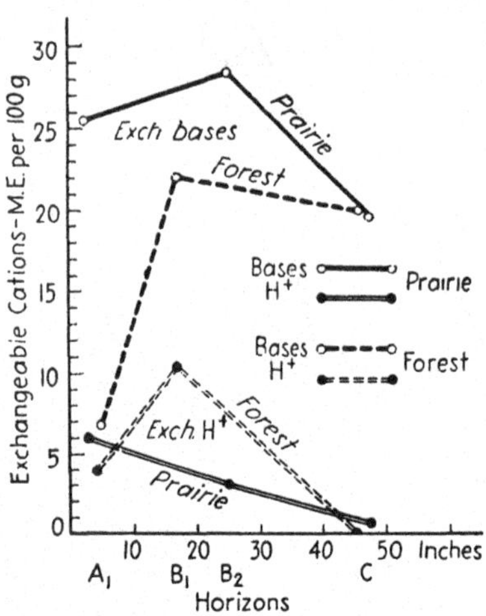

FIG. 111.—Comparison of soils developed under prairie and under forest (Illinois). All other soil-forming factors are identical.

Zone of Carbonates.—Presumably the original loess mantle was calcareous throughout. Leaching has removed the carbonates, and they are now far below the surface, at 52 in. in the prairie soil and at 67 in. in the timber soil. The difference of 15 in. must be attributed to an accelerated movement of carbonates under the influence of forests.

Acidity, pH.—From Fig. 110 it may be seen that in all horizons the pH is lower under timber than under prairie. The lower *A* horizons and the upper *B* horizons are highest in acidity. Near the zone of carbonates, the reaction becomes alkaline.

Exchangeable Bases and Degree of Saturation (Fig. 111).—The sum of the exchangeable bases and the exchangeable hydrogen ions is designated as cation exchange capacity. For any soil horizon, it is primarily a function of the clay and humus colloids and therefore tends to be high in the zone of clay accumulation.

In the forest profile, it amounts to nearly 35 milliequivalents of cations per 100 g. of soil (B_1 horizon, not shown in Fig. 111). The amount of exchangeable bases (Ca + Mg + K + Na) and the degree of saturation $\frac{\text{exchangeable bases}}{\text{cation exchange capacity}} \cdot 100$ are invariably higher in the grassland profile. In other words, prairie vegetation plays the role of a base conserver.

In *conclusion*, it may be said that the Illinois data are corroborated by those from Minnesota insofar as they both reveal a higher degree of translocation of soil materials under timber than under prairie. The Illinois soil types have had a longer period of formation, and therefore the profile characteristics are more distinct than in the Minnesota series. To the soil surveyor, the most striking difference between the two vegetational types is the darker color of the prairie profile, which is the result of a higher content of organic matter. Aside from this, the horizon differentiation is a matter of degree rather than kind. Bray states that in still later stages of development the differences caused by vegetation become very slight.

Relation of Chemical Composition of Plants to Profile Features.—Most of the rain water that enters a forest profile must first pass through a layer of litter or partly decomposed leafmold. The percolating water dissolves various substances from the organic cover, and the profile features are influenced accordingly. At the Central Forest Experiment Station in Moscow, U.S.S.R., Stepanov (59) has made some interesting studies on the decomposition and leaching of forest litter derived from various plant species. A series of lysimeters was filled with leaves and embedded in the ground under the trees from which the leaves were obtained. The containers remained in the forest for 11 months (1927–1928) and received only the natural rainfall and snow. During wintertime, the material was frozen solid. After each rain or period of thawing, the percolated water was collected and analyzed (Table 61).

In view of the discussion of forest vegetation in relation to profile characteristics, the data in Table 61 are exceedingly illuminating, in that they emphasize the significance of the ground vegetation. The leaves of the *shrubs* are richest in mineral constituents, decompose most rapidly and completely, and yield neutral or alkaline leachates. These plant materials may be classified as being inherently of the mull-forming type. The *trees* for the most part are lower in ash; some of them yield percolates of high acidity, and many decompose so slowly that they produce raw humus at even moderate temperatures.

TABLE 61.—COMPOSITION AND REACTION OF LEACHATE OF LEAVES (*Stepanov*)
(Arranged according to average pH of leachate)

Species	Per cent ash, (oven-dry leaves)	Per cent $CaO + MgO$, (oven-dry leaves)	pH of leachate	
			Average	Extremes
Sambucus racemosa (elderberry)	13.28	5.35	7.43	6.1–8.3
Prunus padus (cherry)	14.09	5.66	7.35	6.7–7.7
Caragana arborescens (Siberian pea tree)	10.36	3.37	7.19	6.9–7.5
Corylus avellana (hazel)	10.58	4.71	7.11	6.2–7.4
Viburnum opulus (cranberry bush)	6.38	2.91	6.99	5.5–7.8
Sorbus aucuparia (ash)	6.04	3.47	7.24	6.6–7.4
Betula verrucosa (birch)	7.89	2.99	6.82	—–7.1
Picea excelsa (spruce)	7.06	2.16	6.44	5.5–7.0
Quercus pedunculata (oak)	8.05	2.27	6.31	5.6–6.9
Acer platanoides (maple)	6.32	2.69	5.73	4.3–6.5
Pinus silvestris (pine)	2.46	1.33	5.71	5.5–6.8

TABLE 62.—COMPOSITION OF LEAVES AND LEAFMOLD OF THE PRACTICALLY VIRGIN FOREST ON STAR ISLAND, MINN.

Forest type	Freshly fallen leaves		Leafmold		
	Ash, per cent	CaO, per cent	Ash, per cent	CaO, per cent	pH
Norway pine	4.27	1.35	42.3	0.85	4.5
White pine	4.33	1.52	43.4	1.61	5.1
Maple-basswood	8.44	3.95	48.9	3.84	6.5

The work of Alway (2) and his associates on the nearly virgin forest floor of Star Island, northern Minnesota, furnishes similar data. Here also the leaves of the deciduous trees are richer in minerals and lime than those of the pines, and the pH of the leafmold and decomposed material varies accordingly (15).

Kappen (33), in Germany, treated spruce needles with water (ratio 1:20) and obtained a pH of 3.96 for the solution. Water extracts of *Vaccinium myrtillus* and *Calluna vulgaris* had pH values of 4.8 and 5.9. The hydrogen ion concentrations for the leachates of beech leaves and hazelnut leaves corresponded to pH of 7.1 and 7.2, respectively. Press juice from sphagnum moss had a pH of 3.37.

In 1926 Hesselmann (22) undertook a detailed study of the reaction and lime content of forest litter in relation to profile features. He distinguished five vegetational types according to their content of basic and acid buffer substances.

From the compositions of the leachates given above, it can readily be seen that the nature of the various plant covers must have a profound effect on the type and speed of soil-forming processes. Indeed, many investigators refer to the weathering of rocks that takes place under certain coniferous trees as "acid weathering."

Studies on Forest Floors (A_0 Horizons).—Bodman (7) made a comparison of soils developed under mature stands of pine and of fir in a mixed forest of the Sierra Nevada. This forest developed on Tertiary and Quaternary, gray basalt in a climate characterized by dry summers and wet winters (annual *PE* index = 76.1; *TE* index = 42). On identical types of topography, stands of *Pinus ponderosa* and *Abies concolor* are scattered at random within the forest and thus permit evaluation of the individual effects of pine and fir on soil formation, in accordance with Eq. (36).

The basaltic lava has weathered to produce, immediately beneath the forest floor, a lateritic, reddish brown loam to clay loam. This becomes slightly yellower, more finely textured and more compact with depth, the parent rock being reached at 3 to 6 ft. from the surface. No bleached layer is present in spite of the fact that the annual climatic data correspond to those of podsols. Below the first foot of soil, the profiles under pine and fir are much alike in most respects.

The main profile differences are restricted to the forest floor or A_0 horizon. In all parts of the floor, the fir is more nitrogenous than is the pine. The carbon-nitrogen ratio of the fir floor varies from 26.0 to 42.3, whereas the corresponding values for the pine floor are 33.5 to 54.9. Acidity likewise shows marked differences. The average pH of the coarse undecomposed fraction of the floor is 5.49 for the fir and 4.61 for the pine. The corresponding values for the finer, decomposed fractions are 5.91 and 5.06. Other pronounced differences are found in the contents of calcium and phosphorus. The fir floor contains from 0.25 to 0.52 per cent P_2O_5 and from 3.22 to 3.54 per cent CaO. The pine floor is poorer in these constituents (P_2O_5 = 0.18 to 0.38 per cent; CaO = 1.60 to 1.74 per cent).

Degeneration and Regeneration of Profiles under the Influence of Vegetation.—In going from the humid temperate to the humid cold region in Europe, one passes from brown forest soils to podsols. Both are considered by many European investigators as climatic soil types that differ in general in the following features (Table 63).

TABLE 63.—SELECTED FEATURES OF FOREST SOILS

Features	Brown forest soil (Ramann brown earth)	Podsol
Type of forest	Deciduous	Coniferous
Ground cover	Herbs and grasses	*Calluna* and *Vaccinium*
A horizon	Dark, crumb mull, slightly acid; no typical A_2 horizon	Dark, strongly acid raw humus; typical whitish or ash-gray A_2 horizon
B horizon	Cocoa brown	Rust-brown, occasionally iron hardpan

In southern Sweden, the transition zone between the two soil types occupies considerable territory. In this zone, the brown forest soils and the podsols occur in close proximity. Tamm has discovered that the nature of the *plant cover* is an outstanding factor in the distribution of the two types. One might speak of a metastable soil equilibrium and formulate it as follows:

$$\text{Brown forest soil} \underset{\text{(Vegetation)}}{\rightleftarrows} \text{Podsol}$$

Vegetation decides the trend of the "reaction." From the viewpoint of practical forestry, the brown forest soil is superior to the podsol, and for this reason Tamm (61) designates the left-right process as *degeneration* and the right-left reaction as *regeneration*.

Since Tamm's conclusions are based on many thousands of detailed profile examinations, some of his observations will be discussed in the following paragraphs.

Degeneration of Brown Forest Soils.—The extreme southern part of Sweden has a humid temperate climate corresponding to an annual precipitation of from 16 to 25 in. and a mean annual temperature of from 43 to 45°F. Beech and oak forest are the natural vegetation, and the regional profile is a typical brown forest soil that occurs independently of parent material. However, when the deciduous forest is replaced by coniferous forest or by Calluna heath—which often happens under the influence of man—the brown forest soil degenerates; it becomes strongly podsolized, and true podsols are developed. All characteristic podsol features, such as raw humus, high acidity, etc., are observed.

Regeneration of Brown Forest Soils.—In the western portion of southern Sweden, the rainfall varies between 27 and 39 in., and, although a deciduous beech forest still represents the climax association, the ground-cover vegetation is poorly developed, and a strong tendency for raw-humus accumulations exists. Podsolization is the predominating soil-forming process. In these districts, Tamm inspected a birch forest that had replaced the original beech vegetation by natural immigration into pine plantations. In contrast to the surrounding beech forests that overlay podsolic soils, the ground vegetation of the birches consists of a rich carpet of herbs, grasses, and dwarf shrubs. Under the influence of this new vegetation, a second soil-forming process has been instigated. The acidity has dropped from pH 4 to pH 5.5, and the raw humus and the leached A_2 horizon have been converted into a mull, probably with the aid of worms and other lower animals.

Evolution of Soil Profiles in New England.—A striking example of soil regeneration—the term is used in the utilitarian sense—is presented by the Harvard Forest in New England. When the pioneers arrived in Massachusetts, they found a virgin, mixed forest of indeterminate past. The land, after being cleared, was tilled and pastured for many generations. At Petersham, which is the site of this study, the land was abandoned after 100 years of cultivation and subsequently was occupied by white pine for a period of about 80 years. Following this period, a gradual infiltration of young hardwoods took place. For the past 40 years the vegetation has consisted of a mixed hardwood forest which, it is believed, will ultimately revert to the regional primeval forest. This development series can be summed up as follows:

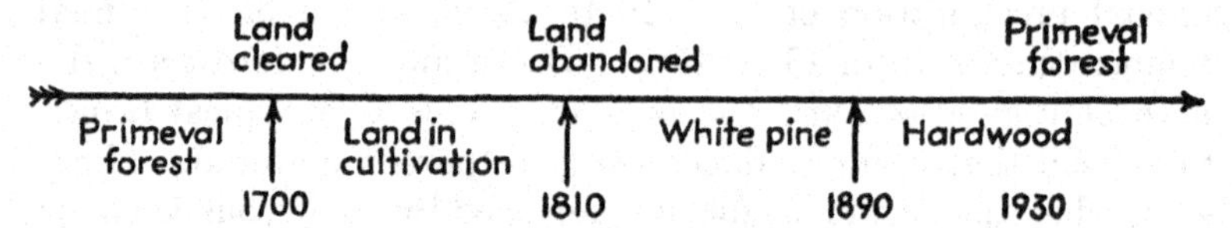

The soil profile of the undisturbed virgin forest of the white pine-hemlock type found in the vicinity of Petersham possesses podsolic features. The cultivation practices that were applied to the cleared land removed all traces of horizons: thus the invading pines took root in undifferentiated soil profiles. Griffith, Hartwell, and Shaw (20) have investigated the soil changes that were brought about by the white pines and the succeeding hard-

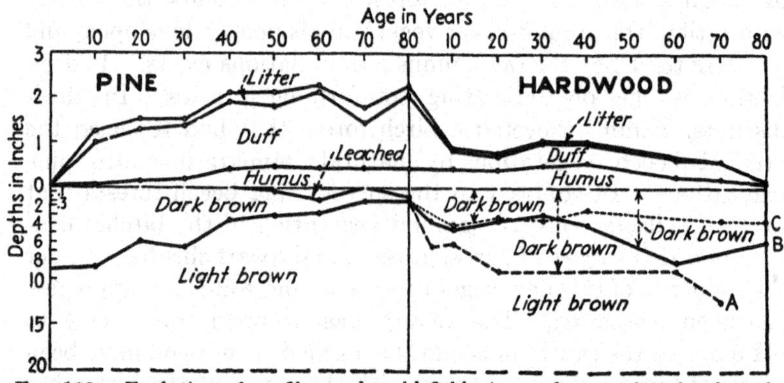

FIG. 112.—Evolution of profiles under old field pine and succeeding hardwood. (*Griffith, et al.*)

woods that consisted mainly of oak, beech, and chestnut. By selecting young and old sites, they were able to evaluate the changes as a function of time. Some of their results are shown graphically in Fig. 112. The soil-profile evolution is marked by the following features:

Soil Changes under Pine Succession

Ten-year Age Class of Pines.—These profiles still reflect the influence of the cultivated fields upon which the pine stands developed. The litter zone is shallow, the duff zone is generally less than 1 in. deep, and there is practically no true humus. The dark-brown zone is deep, averaging 9 in. in thickness. This is about the depth of the old cultivated horizon before the occupancy of the pine. In the graph, the bottom of the enriched horizon is denoted by the broken line at 20 in. where the *C* horizon begins.

Twenty-year Age Class of Pines.—The profiles show an increased amount of duff or raw humus and a small amount of true humus. The dark-brown zone has narrowed to an average thickness of about 6 in.

Sixty-year Age Class of Pines.—The profiles are marked by a deeper organic horizon and a higher proportion of raw humus. Seventeen out of thirty-three profiles show a trace of leaching.

Eighty-year Age Class of Pines.—There is a further decrease in the thickness of the dark-brown layer. It has now an average thickness of less than 1.5 in.

Soil Changes under the Hardwood Succession

Ten-year Age Class.—During the first 10 years of the hardwood occupation, the organic layers that had accumulated under the pine disappeared. Most of the current hardwood litter decomposes annually, and the entire organic horizon is less than 1 in. deep. The dark horizon has deepened considerably.

Old-age Classes.—The organic horizon continues to remain a shallow layer, generally less than 1 in. deep, with practically no litter and with about equal amounts of duff and humus. The thickness of the dark horizon fluctuates considerably. In order to bring out the relationship between this fluctuation and the composition of the prevailing leaf litter the plots were segregated into three classes, *a*, *b*, and *c*, which are based upon susceptibility of the leaves to decomposition.

a. Ash, elm, basswood, white birch, yellow birch,

b. Hornbeam, pin cherry, black cherry, aspen, black birch, soft maple, hard maple,

c. Red oak, white oak, beech, white pine, chestnut.

The average depth of the brown horizon is greatest in the *a* class and smallest in the *c* class.

In conclusion, it may be said that pine develops a deep duff zone of felted needles. The arrested decomposition of organic debris limits the thickness of the dark-brown horizon through lack of infiltration of humus colloids. This degenerating process is reversed with the intrusion of hardwood. The rapid decomposition of the hardwood forest litter increases the thickness of the dark-brown zone below the surface.

General Remarks.—It is well to point out that the soils discussed by Tamm and Griffith, *et al.*, are still young as measured in years, and the majority of the primary minerals have undergone only surface weathering. It takes but little translocation of weathering products to produce either a podsol or a brown forest soil. Thus repeated soil degenerations and regenerations are entirely possible. Nevertheless, it should be remembered that in humid regions the leaching action of percolating rain water is essentially unidirectional and continuous throughout all vegetational successions. In the course of time, leaching will impoverish the soil in plant food to such an extent that regeneration processes will become more and more difficult.

C. MAN AS A SOIL-FORMING FACTOR

Man's activities in relation to soils are many sided (58). In his agricultural endeavors, he modifies effectively the soil-forming factors, notably the vegetational environment. By means of irrigation, he may change completely the climate of the soil. Man also influences directly the properties of the soil. Cultivation and fertilization of soils and the removal of crops are widely practiced activities that stamp man as an outstanding biological soil-forming factor. In the following sections, we shall undertake to discuss a number of selected topics that have a special bearing on agricultural practices.

1. Influences of Cultural Practices

Effect of Cropping on Soils.—An exhaustive exposition of soil changes under the influence of cultivation practices is of course impossible within the space allotted in this book. As in preceding chapters, we must be content with a rather arbitrary selection of data that serve to amplify the quantitative method chosen in this treatise.

Nutrient Losses in Closed Systems.—In 1915, Burd and Martin (11) initiated a series of observations of the effects of cropping upon the composition of 13 soils from California. All soils had been cultivated previously. They were brought to Berkeley,

Calif., and placed in open galvanized-iron containers 60 in. long, 30 in. wide, and 18 in. deep. The containers were suitably arranged for subirrigation and insulated against lateral temperature changes by an external boxing filled with field soil.

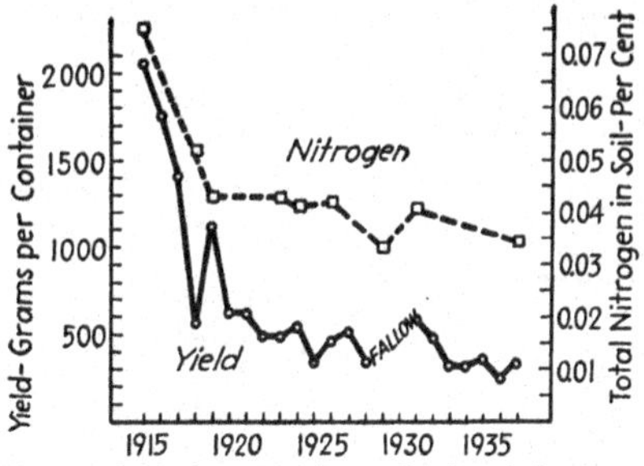

FIG. 113.—Fluctuations of barley yields and of total nitrogen content of soil of Burd and Martin's tank experiments.

The soils were cropped to Beldi barley. Distilled water exclusively was used for irrigation, the tanks being covered with watertight canvas covers during rainstorms and when precipitation was anticipated. The crop yields and total nitrogen fluctuations for soil 8*A*, a Fresno fine sandy loam from San Joaquin Valley, are shown in Fig. 113.

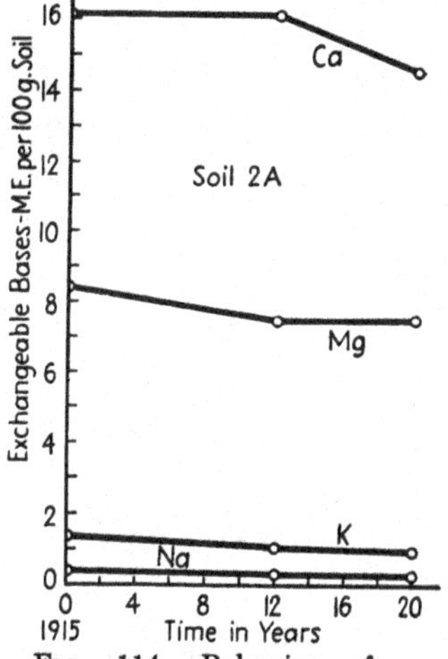

FIG. 114.—Behavior of exchangeable bases in the tank experiments of Burd and Martin.

The most impressive features of the curves are the rapid decline of yield and of soil nitrogen during the early periods and the damped oscillations in the later years. Today, crop yield and nitrogen are still declining, but at an exceedingly small rate. It will take at least another decade to learn whether or not a stationary state is being reached. Owing to the fact that the experiment was so designed that no leaching can take place, an accurate balance sheet of nitrogen economy can be kept. For the period of greatest decline in yield (1915–1919), the 1,600 lb. of soil 8*A* lost 211 g. of nitrogen; however, since the crops withdrew a total of only 101 g. of nitrogen, there was an absolute deficiency of 110 g. of nitrogen. Most likely, this loss was due mainly to biological reduction in the soil itself. For the 13 soils tested, the average extra loss per season to a depth of 9 in. amounted to about 100 lb. per acre. Similar losses have been reported by Lipman and Blair for soils in New Jersey (38).

TABLE 64.—BARLEY YIELDS AND NITROGEN OF SOIL 2A

Year	Soil 2*A*, Yolo silty clay loam	
	Yield, grams per tank	Total soil nitrogen, per cent
1915	1,957	0.137
1927	564	0.109
1935	511	0.109
Reduction in 20 years, per cent	74.0	20.4

Burd and Martin also have analyzed the aforementioned California soils for mineral constituents. The amounts of exchangeable cations, which represent a significant reservoir of easily available plant nutrients, are given in Fig. 114 for soil 2*A*. Compared with the sharp declines in crop yields and in nitrogen of this soil (see Table 64), the reduction in total exchangeable bases is small, amounting to only 10 per cent. These results appear to be characteristic of many irrigated soils of arid regions. Nitrogen rather than the mineral complex is the first limiting factor in crop production. However, if liberal amounts of nitrogen are provided, intensive cropping will soon affect the base status of the soil. Hoagland and Martin (24) grew five barley and five tomato crops successively on 20 kg. of soil to which had been added sufficient amounts of nitrogen and phosphorus to ensure satisfactory growth. Figure 115 gives a clear

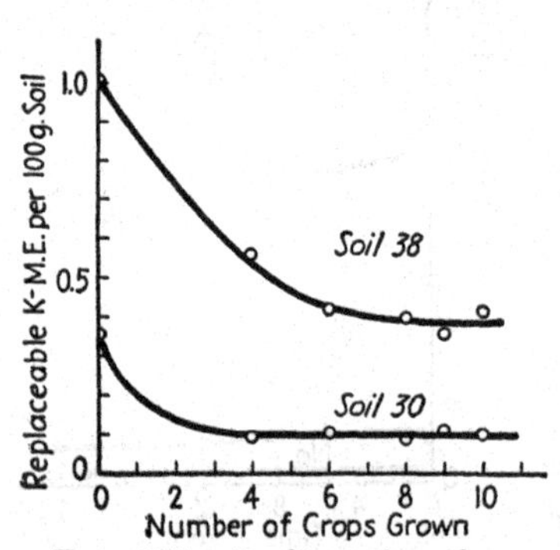

FIG. 115.—Decline of exchangeable potassium under conditions of intensive cropping.

picture of the drastic effect of the intensity of cropping on the level of exchangeable potassium in the soil. In the cases of soils 30 and 38, an apparently stable potassium content was reached after four and six crops had been harvested. Since the variations in yield were slight, the mineral mass of these soils apparently has the power of supplying a relatively steady flow of exchangeable potassium. It is not known for how long a period the two soils are capable of maintaining the observed exchangeable potassium levels of 0.39 and 0.10 milliequivalents per 100 g. of soils.

As a broader aspect of the results of controlled experiments, we observe that neither the yield nor the nutrient status declines at equal rates. Sooner or later a stationary state is reached that permits a steady though low production of crops.

Nutrient Losses in Prairie Soils.—Some 15 years before the instigation of the recent intensive research in soil erosion, Miller and Duley (44) initiated the first scientific, long-time experiments on the destructive effects of this process. For moderate slopes, the losses of soil fertility were negligible under grassland cover but of appalling proportions under customary cultivation practices. Soon the question arose as to the dimensions of the fertility losses under ordinary cropping methods in the absence of erosion. In 1929, M. F. Miller, Dean of the Missouri College of Agriculture, suggested to the author a study of fertility changes in level prairie soils where erosion is of negligible influence.

On the prairie lands of Callaway County, Missouri, there exists a piece of land of about 100 acres area that has never been plowed. The adjoining cultivated fields, originally also prairie, so strongly resemble the virgin tract in topography and profile features that it appears reasonable to assume that both prairie and cultivated lands originally were alike. The entire physiographic unit known as the "Grand Prairie" is composed of Putnam silt loam, a loess profile that is characterized by a dark-gray colored surface soil overlying a heavy subsoil stratum. The topography is nearly flat, and, consequently, losses of soil due to erosion are reduced to a minimum.

The field designated in this study as prairie is not strictly a virgin prairie in its floristic meaning, because for many years it has been pastured and cut for hay. Bluestem (*Andropogon furcatus*), a typical prairie plant, is the predominating grass.

One of the adjoining fields was brought into cultivation shortly after the Civil War (1865) and has been cropped ever since. The main crops grown have been corn, oats, and wheat. In 1910, cow peas were raised. Corn has been the prevailing crop since 1906. No artificial fertilizers, lime, or stable manure have ever been applied. This system of soil management corresponds to the customary farming practices of that region.

From the cultivated field, 75 surface-soil samples to a depth of from 0 to 7 in. were collected and analyzed individually. The same number of samples was taken from the prairie, except

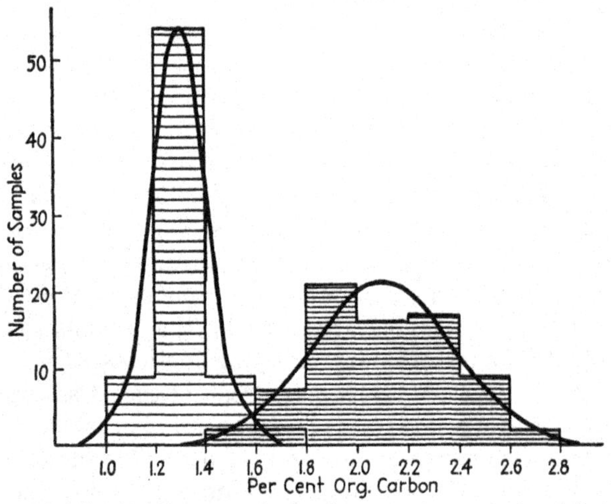

FIG. 116.—Distribution diagrams of organic-matter content of prairie soil (dark) and cultivated field (light).

that the depth interval from 1 to 8 in. was selected. The first inch of soil, consisting of partially decomposed leaves and stems, was discarded.

Reduction of Total Organic Matter.—The average content of organic carbon (C) was 2.10 ± 0.032 per cent for the prairie soils and 1.30 ± 0.014 per cent for the cultivated soils. Taking into consideration the values of the mean errors, the difference 0.80 ± 0.035 per cent C is highly significant. In terms of the original value of the prairie, cultivation had lowered the organic-carbon content by 38.1 per cent. Multiplying C by the conventional factor 1.742 gave the total organic matter (humus) which was 3.66 per cent for the prairie and 2.26 per cent for the cultivated field. In other words, about 28,000 lb. of humus per acre had been oxidized during the course of 60 years.

It is of interest to note that cultivation not only affected the average carbon content but also changed the variability of the *A* horizon. Figure 116 shows two types of distribution curves. First the zigzag lines, indicating the frequency distribution for class intervals of 0.20 per cent C, and, second, the smoothed curves, calculated from the probability equation of Gauss (compare page 29). The parameter h, which corresponds to the maximum height of the curve, is a numerical index of soil uniformity. The greater the value of h, the more uniform, or less variable, is the field. For the prairie, h amounted to 0.518; the corresponding value for the cultivated field was 1.193. Both the graph and the h values conclusively show that cultivation had rendered the topsoil areally more uniform.

Loss in Total Nitrogen.—The nitrogen content of the prairie was 0.197 ± 0.0027 per cent nitrogen, whereas the cultivated field contained 0.129 ± 0.0013 per cent nitrogen. The absolute difference between the two levels was 0.068 ± 0.003 per cent nitrogen. The percentage reduction in nitrogen over 60 years of cultivation was slightly less than in organic matter, namely, 34.5 per cent, or about one third. The prairie had over 50 per cent more total nitrogen than the cultivated field. Cropping had increased the uniformity of the soil also with respect to nitrogen as indicated by the h parameters, the magnitudes of which were 0.610 for the prairie and 1.297 for the plowed area.

Change in Carbon-nitrogen Ratio.—On the basis of the aforementioned arithmetical averages, the carbon-nitrogen ratios—which play an important role in many soil-fertility aspects—had the following values:

$$\begin{aligned} \text{Prairie} &= 10.66 \pm 0.097 \\ \text{Cultivated field} &= 10.08 \pm 0.041 \\ \text{Difference} &= 0.58 \pm 0.105 \text{ (mean error)} \end{aligned}$$

Cultivation had brought about a narrowing of the carbon-nitrogen quotient which—though small (5.4 per cent)—was statistically significant.

Increase in Soil Acidity.—In all samples, the hydrogen ion concentration was determined with the aid of the quinhydrone electrode. The arithmetical average of the pH values is 5.34 ± 0.025 for the prairie and 5.01 ± 0.012 for the cultivated field. Clearly, cultivation and cropping have increased soil acidity. In terms of absolute hydrogen ion concentrations, the pH difference of 0.33 corresponds to an increase of 114 per cent in actual soil acidity.

Removal of Mineral Constituents.—The readily available nutrient cations were determined by adding 0.10 normal hydrochloric acid to the soil. The amount of hydrochloric acid not

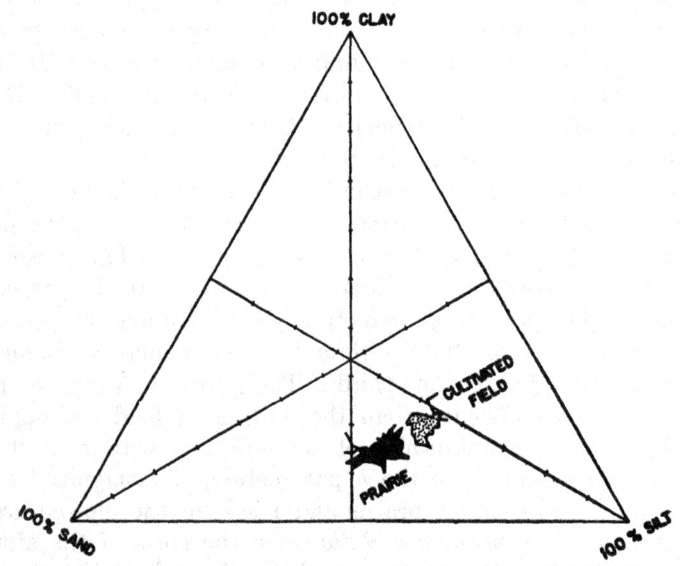

FIG. 117.—Mechanical analyses of prairie soil and cultivated field (73 samples each). Note the general shift from the coarser to the finer grains.

neutralized by the bases of the soil was ascertained by back titration with sodium hydroxide. The ions thus determined are related to the exchangeable bases of this soil. For the prairie soil, the base status was found to be 10.6 ± 0.20 milliequivalents per 100 g., whereas for the cultivated field the corresponding value was 8.0 ± 0.22. The decline amounts to 25 per cent of exchangeable bases in the surface soil. In terms of limestone material, this loss corresponds to about 2,600 lb. per acre.

Alterations of Soil Structure.—In the virgin prairie, the colloidal clay particles are aggregated to larger units that have the size of silt and sand grains. Evidence points strongly to the fact that organic matter acts as a binding agent between the particles. Therefore one might anticipate that destruction of organic matter brought about by cultivation would be accompanied by changes in the aggregation of the clay particles and the structural pattern of the soil. Bouyoucos' method of mechanical analysis of soil—in modified form—was used to determine in an arbitrary manner the changes in the state of aggregation of the soil particles. No dispersing agent was added to the material. The triangular presentation of Fig. 117 shows convincingly that cultivation tends to shift the distribution of aggregate sizes in the direction of smaller units. In the prairie, 39 per cent of the soil mass exists in the form of particles of the size of sand. Cultivation reduced this value to only 28 per cent, which corresponds to a reduction of nearly 30 per cent. Correspondingly, the "clay fraction" is increased from 14 per cent in the prairie to 19.5 per cent in the cultivated field. This difference amounts to nearly 40 per cent. These alterations in soil structure bring about soil compaction that hampers air and water circulation as well as tillage operations. Conditions for plant growth and soil management are thus made less favorable.

Effect of Burning.—The possible utilization of thousands of acres of the poorer sand lands in the southeastern Coastal Plains as timberland has focused attention on the problem of protected reforestation. One of the greatest obstacles in the utilization of these poorer lands, however, has been the yearly burning of forests and cutover lands in an effort to produce sufficient grass for the maintenance of range cattle. For many years, the forestry service has been emphasizing the resultant injury to tree growth and to the possible future value of the timber. About 10 years ago, Barnette and Hester (3) gathered quantitative information on the effect of burning on soil properties. On Norfolk medium-fine sand in Florida, they found an area of virgin pine forest that had not been burned for the previous 42 years. They compared it with an area on a similar type of soil that had been burned over almost yearly for the previous 42 years. Barnette and Hester's findings are reported in Table 65. The undisturbed forest is distinctly richer in organic matter and nitrogen than the burned areas. The difference is the more important, since nitrogen is usually one of the limiting elements in soils of southern regions.

The burned area tends to be slightly less acid and, in the surface layer, contains more exchangeable calcium than the virgin forest, a condition that is no doubt due to the accumulation of ash in the surface soil. A significant condition, not brought out in Table 65, is the existence of a 4-in. layer of pine needles and leafmold above the mineral soil of the virgin forest, which is entirely missing in the burned area. This forest floor contains from 59 to 96 per cent organic matter and from 0.60 to 0.72 per cent total nitrogen. Accurate measurements of the weight of the forest floor indicate that, on the average, burning has destroyed annually 2,888 lb. of organic matter and 27 lb. of nitrogen per acre.

TABLE 65.—ANALYSES OF FOREST SOILS FROM BURNED AND UNBURNED AREAS (*Barnette and Hester*)

Constituents	Treatment	Depth, inches (mineral soil)			
		0–9	9–21	21–33	33–45
Organic matter, per cent	Unburned area	2.54	1.34	0.97	0.61
	Burned area	1.35	0.61	0.45	0.37
Total nitrogen, per cent	Unburned area	0.042	0.021	0.014	0.015
	Burned area	0.025	0.015	0.014	0.012
pH	Unburned area	5.1	5.5	5.8	6.1
	Burned area	5.7	5.8	5.9	5.7
Replaceable Ca, milliequivalents per 100 g. of soil	Unburned area	0.73	1.00	1.13	0.88
	Burned area	1.07	0.77	0.64	0.68

The importance of pine as a means of increasing the organic-matter supply and the ultimate fertility of the Norfolk sands cannot be overemphasized, and the effects of burning over these lands distinctly interferes with the accumulation of organic matter in the soil.

Changes of Soils under Irrigation Practices.—In arid regions, intensive agriculture must rely on irrigation. This form of human enterprise sets into action an entirely new series of soil processes that require careful observations and control lest the productivity of the soil be impaired beyond economic recovery. Where the irrigation water is pure and the ground-water table is low, the resulting changes are similar to those brought about by leaching in semihumid and humid regions. Frequently, however, the irrigation water contains considerable amounts of salt. In these cases, the crops remove the moisture of the soil, and the salts accumulate in the soil solution. This process may result in the formation of alkali soils.

Salinization.—A striking example of artificial salinization was observed at Station 2 of the Salt River Valley Irrigation Project in Arizona. The salt content of the irrigation water varied between 800 and 2,500 parts per million and consisted mainly of calcium, sodium, chloride, sulphate, and bicarbonate. The rate of application of irrigation water was about 3 acre-ft. per season.

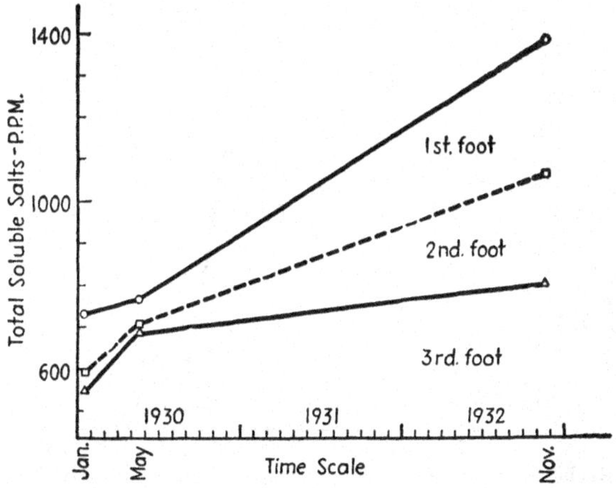

FIG. 118.—Augmentation of salt content of an Arizona soil as a result of irrigation.

Soon radical changes in the salt content of the various soil layers were observed (Fig. 118). Prior to irrigation, the soil contained from 553 to 733 parts per million total soluble salts; after 3 years, the salinity had risen to nearly 1,400 parts per million in the surface layer. The lower parts of the profile were less affected, but even in the third foot the salt concentration was raised by nearly 50 per cent. Absence of salt in the irrigation water does not necessarily guarantee immunity from alkali injury. Heavy irrigation tends to raise the ground-water level, and, where the latter is saline, injurious salts may reach the root zone by capillary rise. Large areas in the San Joaquin Valley of California have been made worthless for agricultural use by this process.

Reclamation of Alkali Soils.—In 1920, Kelley (34) undertook the reclamation of a black alkali soil near Fresno in the San

Joaquin Valley of California. The soil, classified as Fresno fine sandy loam, contained considerable quantities of soluble salts that consisted chiefly of sodium chloride, sodium sulfate, and sodium carbonate. In addition, the soil contained significant amounts of adsorbed sodium. A large part of the area used for experimentation contained so much alkali as to be highly toxic to crops. Barley, harvested as hay, yielded only 616 lb. per acre. As a first step in the reclamation process, pumping equipment was installed in order to lower the ground-water table and

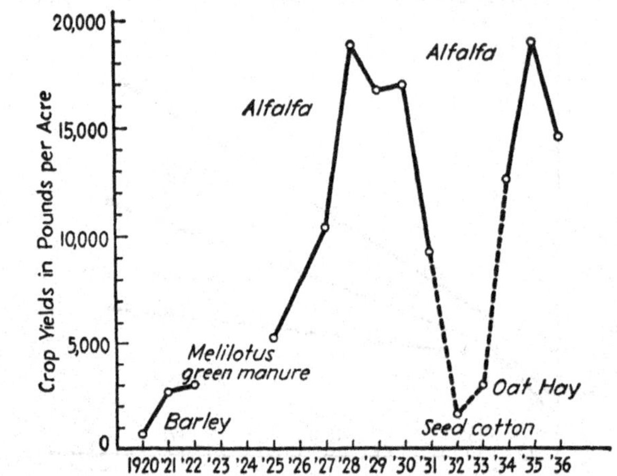

FIG. 119.—Reclamation of alkali soil. Plot 4, Kearny Vineyard Experiment Field, Fresno, Calif.

provide good drainage conditions. In 1920, gypsum was added in the amount of 9 tons per acre. After plowing, the plots were flooded for a period of several weeks for the purpose of leaching out the soluble salts and of carrying the gypsum down into the subsoil. Barley was planted in December. The results showed that the gypsum produced a considerable effect on the soil, but the yields remained unsatisfactory. Chemical tests of the soil revealed that it still contained a considerable amount of sodium carbonate. Consequently an additional 6 tons of gypsum per acre was applied in the summer of 1921. Subsequently, no further material has been added. The land has been cultivated, irrigated, and cropped as indicated in Fig. 119. The gypsum produced moderate increases in the yield of barley hay in 1921 and 1922. During the years 1923 and 1924 *Melilotus indica* and

M. alba were grown and plowed under as green manure. Subsequent plantings of alfalfa produced 5,272 lb. of hay in 1925. In 1926, the land remained uncropped. Alfalfa was again planted during the years 1927 to 1931, and the progressive increases in yield emphatically portray the effectiveness of the reclamation of the soil. Soil analyses indicate that practically all the soluble salts and the greater part of the adsorbed sodium have been removed. The poor physical condition of the alkali soil, characterized by harsh clods, was successfully overcome, and the soil became granular and mellow. The yields of 1931, which were lower compared with those of previous years, were not caused by alkali but by an invasion of Bermuda grass and a severe infestation of alfalfa wilt. After a rotation that included cotton and oats, alfalfa again was planted in 1934. The yields were satisfactory and remained so throughout the subsequent years, demonstrating the success of the reclamation project.

2. Concepts of Soil Productivity and Soil Fertility

Changes in soil properties are designated by agriculturists as alterations of soil productivity and soil fertility. It will prove of interest to examine these cardinal agronomic terms in their relationship to soil properties and soil-forming factors.

When a farmer buys a tract of land, the surrounding climate is implicitly included in the transaction. In reality, he acquires a *producing system* of which he may become a part. The physical productivity of this producing system is measured in yields per unit area, be it tons of hay per acre, or bushels of oats per acre.

For the purpose of a quantitative elucidation, it is customary to divide the producing system into a number of components, namely, climate, soil, plant, and man. The magnitude of the yield depends on these factors and on the growth factor time. A number of investigators have stated this relationship in the form of an equation as follows:

$$\text{Yield} = f\,(\text{climate, plant, man, } \textit{soil}\text{, time}) \qquad (37)$$

The equation indicates that the amount and kind of organic matter harvested on a given area depend on the factors in parentheses. This type of formulation has been the source of much controversy and confusion among agriculturists. Some accord the factors of the right-hand part of the equation the rank of independent variables, whereas others object to such treatment, contending that soil is not an independent variable because its properties are conditioned by these very factors. Where, then, is the fallacy? The difficulty is readily overcome when we realize that instead of "soil" in Eq. (37) we should have written "parent material" or "soil material." For, as was pointed out in previous discussions, any change in one of the soil-forming factors starts a new cycle of soil formation. The introduction of man's activities and the elimination of natural vegetation alter the effective biotic factor, and the body hitherto called "soil" automatically assumes the rank of parent material or soil material, which has been defined as the initial state of a new process of soil development. It is for this reason that the word "soil" in Eq. (37) is written in italics. This indicates that the agriculturist's use of the word "soil" differs from the more rigorous concept employed by pedologists. Keeping this logical distinction in mind we come to realize that Eq. (37) is identical with the fundamental equation of soil-forming factors. Soil properties of the s type as well as the yields are truly dependent variables, whereas all factors within parentheses may be treated as independent variables. These latter may be designated as productivity factors. Each contributes a certain share in the production of a given quantity of organic matter.

Definition of Soil Productivity.—Difficulties immediately arise when we speak of *soil productivity*. Soil is only one component of the above equation; it has a much more restricted meaning than land. We may readily accept yield as the numerical index of the productivity of the general producing system, but are we justified in choosing this same yield as a measure of soil productivity, which is only one of the production factors?

When the productivities of two given areas are compared, it must first be determined whether or not the difference in yield is due entirely to the difference in soil. A comparison of the cotton yield of a Cecil sandy loam in Alabama with the corn yield of a Tama silt loam in Iowa deals with *producing systems* rather than with soils, because all factors in Eq. 36 have different magnitudes. On the other hand, if, in Saline County, Missouri, we contrast corn yields of experimental fields on the Marshall silt loam with those on the Summit silt loam, we are actually comparing soil productivities, because the determining variables, climate, plant, time, and management, are practically identical. In this case, we are solving the equation

$$\text{Productivity} = \text{yield} = y = f(S)_{cl,v,h,t} \tag{38}$$

where S indicates soil, cl climate, v the crop, h management, and t time. The productivities of the aforementioned Cecil sandy loam in Alabama and the Tama silt loam in Iowa could be accurately compared only under identical conditions of climate, crop, and management. Such studies as yet have not been undertaken on any extensive scale, and, for this reason, we do not know where in the world, or even in the United States, the most productive soils are situated.

For the purpose of defining productivity in quantitative terms, let us assume for the moment that in a given standard system that satisfies Eq. (38), soil A produces 50 bu. of corn per acre and soil B 25 bu. per acre. Obviously in this particular system soil A is more productive than soil B, by 25 bu. However, soil A is not twice as productive as soil B, because the other factors, particularly climate, contributed their share in producing the absolute yields.

As an arbitrary standard or reference system, let us choose the Davis growth chamber, a greenhouse that permits perfect control of light, temperature, and moisture. Let us then introduce into the system—in absence of any soil—a number of seed potatoes. They will germinate and build up organic matter through photosynthesis so that after a given length of time the yield of the system will be y_0 pounds. If we modify slightly the above procedure by introducing 1 lb. of soil material along with the seed potatoes the yield will be higher and will amount to y lb. Since no factor but soil has changed, the entire yield increase may be attributed to the soil material itself. We designate the ratio

$$\frac{\text{Increase in yield}}{\text{Amount of soil added}} = \frac{y - y_0}{1}$$

as the productivity of 1 lb. of soil material in the chosen standard system. Since it is customary to denote numerical differences by the Greek letter Δ (delta), we may write

$$\text{Soil productivity} = \frac{\Delta y}{\Delta S}$$

Naturally, we may conduct the experiment with variable amounts of soil. If the yields thus obtained are plotted against the corresponding additions of soil, the slope at any point is the differential coefficient dy/dS. It represents a measure of soil productivity and is here called the soil-productivity coefficient. To illustrate its meaning a set of data taken from a paper by Jenny and Cowan (31) is reproduced in Fig. 120. Soybean plants were grown in artificial soil material consisting of inert quartz sand

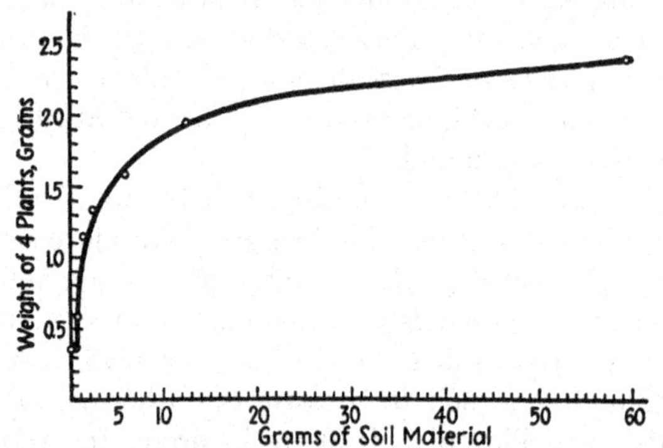

FIG. 120.—Illustration of the growth of four soybean plants as a function of amount of soil material (Ca-clay).

mixed with clay that contained adsorbed nutrient ions. Except for the initial section of the curve, the total yield rises as the amount of soil increases. The productivity coefficient corresponds to the tangent drawn at any point on the curve. It is evident that soil productivity depends not only on the kind of soil but also on the amount.

Under most greenhouse and field conditions, not dy/dS but $\Delta y/\Delta S$ is measured. In the case of the two corn soils A and B, which yielded 50 and 25 bu. per acre, respectively, the yields y_0 produced in absence of soil must be subtracted, and we obtain the following productivities;

$$\text{Productivity of soil } A = y_A - y_0 = 50 - y_0$$
$$\text{Productivity of soil } B = y_B - y_0 = 25 - y_0$$

The ratio of the two productivities is not 2, which substantiates the previous statement that soil A is not twice as productive as soil B.

In general agricultural practice, the final yield y is very large compared with the value y_0, so that the difference $y - y_0$ is very nearly equal to y. In other words, for most agricultural purposes the yield that a given soil produces under a given set of conditions may be taken as the numerical expression of the soil's share in the productivity of the system under consideration. This is permissible whenever y_0, the yield in absence of soil, is very small compared to y, the yield in presence of soil. The above definition of productivity need not be restricted to soils. By changing climate, we may speak of temperature, or light productivities. This is most readily seen by writing Eq. (37) in the differential form

$$dy = \left(\frac{\partial y}{\partial cl}\right)_{v,h,S,t} dcl + \left(\frac{\partial y}{\partial v}\right)_{cl,h,S,t} dv + \left(\frac{\partial y}{\partial h}\right)_{cl,v,S,t} dh + \left(\frac{\partial y}{\partial S}\right)_{cl,v,h,t} dS + \left(\frac{\partial y}{\partial t}\right)_{cl,v,h,S} dt \quad (39)$$

The partial derivatives are the productivity coefficients of the determining variables in Eq. (37).

Inherent Productive Capacity of Soils.—Soil productivity, as defined in the preceding section, varies with different crops, climates, and forms of soil management. A number of investigators have striven for more fundamental concepts and have coined expressions such as productive power and inherent productive capacity of soil.

In order to test these notions, it is necessary that the soils that are to be compared be brought into various climates and planted to various crops. If it should turn out that a given soil C always produces a lower yield than another soil D, no matter what the type of crop, then we would be justified in saying that soil D has a higher inherent productive capacity than soil C. For under no circumstances will the yields of soil C equal or exceed those of soil D. Such conditions are seldom encountered. Only extreme cases of very rich and very poor soils would meet these requirements.

An impressive illustration of the dependency of soil productivity on environment is furnished by a soil exchange experiment with sugar cane conducted by Borden (8) at the Hawaiian Sugar Planters' Experiment Station in 1935. Two different soils, designated herein as X and Y, were placed in tubs and planted to sugar cane (variety H109). One pair of X and Y soils was exposed to the climatic conditions of Makiki, another pair was placed at Manoa, the climatic conditions of which are more humid and more cloudy than those of Makiki. The experiment lasted 14 months. The amounts of sugar produced by the two soils at the two localities are assembled in Table 66. The yields of both soils were higher at Makiki than at Manoa; in other words, the productivity varied with climate. At Makiki both soils were "good"; whereas at Manoa these very same soils could be classified as "poor."

TABLE 66.—INFLUENCE OF CLIMATIC CONDITIONS IN THE PRODUCTION OF SUGAR CANE

Soils	Sugar produced, pounds	
	At Makiki	At Manoa
X	10.63 ± 0.68	3.37 ± 0.20
Y	12.14 ± 1.79	2.54 ± 0.21

In this connection, the report on the interstate soil-exchange experiments published by LeClerc and Yoder (36) merits special consideration. Soil from California was transported to Kansas, and soil from Kansas was sent to California. In both localities, both soils were planted with identical varieties of wheat. The chemical composition of the crops grown in 1910 indicated the following relationships: In California the wheat produced by the California soil contained more protein than the wheat grown on the soil imported from Kansas. In Kansas, however, the situation was reversed. Here, the Kansas soil was superior to the soil shipped from California. Such reversals of behavior do not speak in favor of inherent productive capacities of soils. As a consequence of the transfer of these soils, not only the yields but also the properties of the soil were altered, a fact brought to light in an interesting study by Lipman and Waynick (37).

Soil Fertility.—Liebig's revolutionary views of plant nutrition added momentum to the theory of soil statics that came into great prominence in subsequent years. Its basic principle rests on the discovery that the removal of a crop depletes the soil in mineral nutrients, particularly potash, phosphorus, and nitrogen. Crop yields inevitably decline unless corresponding amounts of mineral plant foods are returned to the soil. In the latter part of the nineteenth century, the problem of maintenance of soil productivity became synonymous with balancing the input and output of nutrient elements.

Probably because of this historic background the concept of soil fertility, unlike that of soil productivity, is closely related to plant nutrients. For instance, it is often stated that soil erosion removes fertility or that the application of stable manure and mineral salts is beneficial because they restore and augment soil fertility. On the other hand, where excessive cultivation causes the soil to be puddled and subject to poor water penetration and aeration, it is not the fertility but the productivity that is said to suffer. Likewise, according to this view, the reclamation of alkali soils improves primarily the productivity of the soil, since nutrient elements are usually not lacking.

Unlike productivity, the fertility of a soil cannot yet be expressed numerically. Although we may find that a given soil contains 0.20 per cent total nitrogen, 100 parts per million of available phosphorus, and that the calcium-potassium ratio in the exchange complex is 20:1, we do not know how to combine these values into a single number that will serve as an index of soil fertility in the manner that yield represents a quantitative symbol of productivity.

Relationships between Productivity and Fertility.—In many soils of the corn belt and adjacent regions, a fairly consistent proportionaity exists between the nitrogen content and the amounts of available mineral plant nutrients such as potassium, phosphorus, and calcium. Under these specific conditions, total soil nitrogen may serve as a simple numerical measure of soil fertility. For soils having a favorable physical constitution, the equation of soil productivity (page 245) may then be written as follows:

$$\text{Yield} = f\,(\text{nitrogen})_{cl,v,h,t} \tag{40}$$

These functions have been most successfully evaluated for corn (*Zea mays*). By way of illustration (53), data from untreated check plots at Wooster, Ohio, on which corn has been grown continuously since 1894 are presented in Table 67. The total nitrogen content and the yield of corn are computed on a relative basis and are compared at three different periods. The relationship is nearly linear.

TABLE 67.—RELATIVE NITROGEN CONTENT AND YIELD OF CONTINUOUS CORN PLOTS AT WOOSTER, OHIO

Year	Relative nitrogen content of soil	Years	Relative corn yield
	Per cent		Per cent
1896	100.0	1894–1898	100.0
1913	51.6	1911–1915	59.6
1925	41.2	1923–1927	45.6

A more complete solution of Eq. (40) is given in Fig. 121. Each point represents the average corn yield (10- to 20-year periods) of unfertilized check plots from field experiments conducted by the Missouri Agricultural Experiment Station. The 12 experimental fields from which the data of Fig. 121 are taken

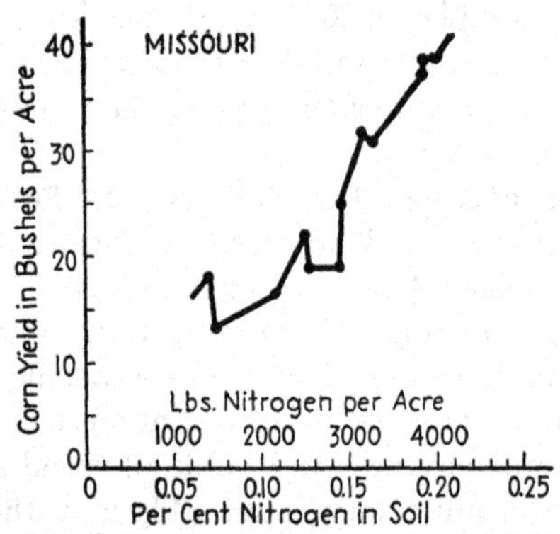

FIG. 121.—Showing the relation between average corn yield and total nitrogen content of soil of Missouri experiment fields.

are located on widely different soil types but have similar climatic environments. It may be clearly seen that, in general, low yields of corn are associated with low contents of soil nitrogen, and high corn yields are found on soils high in nitrogen. A similar type of curve may be constructed for the well-drained upland soils of the state of Illinois (4).

In many greenhouse and field experiments, the nature of the soil material is systematically altered by the addition of fertilizers. The relationship between the yield and the soil ingredient added is usually of the type shown in Fig. 120. In a great number of cases, the functions may be described by the so-called "Mitscherlich" equation:

$$\frac{dy}{ds_n} = k_n(y_{max} - y), \qquad y = y_{max}(1 - e^{-k_n s_n})$$

The symbol y_{max} represents the maximum yield obtainable under the conditions of the experiment, and s_n indicates any soil property that functions as a plant nutrient. It was formerly believed that the value of k_n for a given nutrient element was a universal constant. The principles of agrobiology (68) were founded on this assumption. For details the reader must be referred to current literature on the subject (64).

3. Future Trends of Soil Fertility

In the light of the established functions between soil properties and time that were discussed in Chap. III, it is clearly evident that the formation of a fertile soil rich in nitrogen and organic matter requires periods of hundreds and thousands of years. In contrast, the rapid deterioration of soil fertility under exploitive systems of farming, even in absence of accelerated soil destruction by erosion, unfolds a pessimistic outlook for the future. In the following pages, we shall endeavor to estimate the future trends of soil fertility under conditions where the removal of soil material is negligible, *i.e.*, in areas of level topography or adequate erosion control. The study of the disastrous effects of soil erosion itself is outside of the scope of this book and has been authoritatively dealt with in Bennett's recent book, "Soil Conservation" (5).

Trends of Soil Fertility under Average Farming Conditions.—The study of fertility changes in the Putnam silt loam immediately raises the questions: Do the plant nutrients of the soil continue to decline, and if so, at what rate? For the present, a quantitative answer for conditions in the field can be given only for the element nitrogen. The curve in Fig. 122 is based on data from several Midwestern experiment stations (30) where investigations similar to those on the Putnam silt loam have been conducted. Each point on the curve is the result of a comparison between a cultivated field and the adjoining virgin prairie. The nitrogen content of the virgin prairie is arbitrarily taken as 100. Although the scatter of the points is considerable, a general trend is obvious. At first the rate of decline is marked, but with advancing years the changes become much less drastic. The approximate percentage changes are as follows:

Nitrogen reduction in the first 20 years = 25 per cent,
Nitrogen reduction in the second 20 years = 10 per cent,
Nitrogen reduction in the third 20 years = 7 per cent.

One might anticipate that in the course of time a steady state of the nitrogen content of the soil is reached, during which the removal of soil nitrogen by crops is balanced by natural rejuvenation through nitrogen fixation.

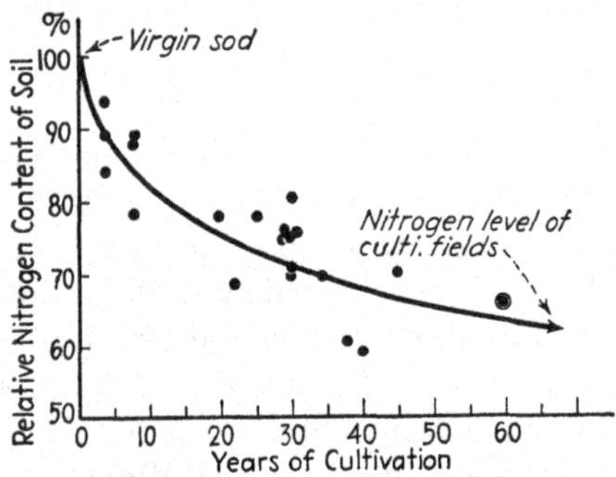

FIG. 122.—Decline of soil nitrogen under average farming conditions in the Central states.

Influence of Cropping Systems.—The nitrogen-time curve shown in Fig. 122 expresses the decline of soil fertility under current farming conditions in the north central United States. It is conceivable that the rates of decline would assume entirely different magnitudes were the cropping and cultivation methods altered. A recent paper by Salter and Green (54) affords ample support for this belief. These investigators have studied the nitrogen and carbon changes in several experiment plots on the Wooster silt loam at Wooster, Ohio. The following cropping systems, which were established in 1894, were studied.

I. Continuous corn,
II. Continuous wheat,
III. Continuous oats,
IV. Five-year rotation: corn, oats, wheat, clover, and timothy,
V. Three-year rotation: corn, wheat, and clover.

There were two treatments: (*a*) no fertilizer, and (*b*) lime. The curves plotted in Fig. 123 refer to soil nitrogen and represent averages for the unlimed and limed plots. The influence of the cropping system is astonishing. Under continuous corn, soil fertility was reduced to about one-half in the brief period of 20 years. The beneficial effect of crop rotations is convincingly brought out, particularly in the case of the 3-year rotation in

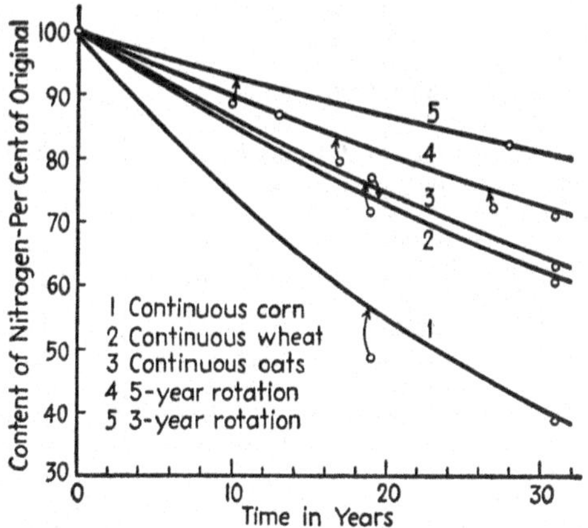

Fig. 123.—Decline of soil nitrogen as influenced by cropping systems (Wooster, Ohio).

which the clover approximately balances the destructive effect of the corn (compare also Table 69).

Similar observations were made on the Sanborn Field at Columbia, Mo. (Table 68). The initial nitrogen content of the soil is not accurately known but probably is in the vicinity of 4,000 lb. per acre. Under rotation systems with heavy applications of stable manure, the nitrogen and organic-matter content apparently are close to the initial value. These data suggest that a soil that has previously been depleted of nitrogen may be rebuilt to a certain extent by the liberal application of barnyard manure and by proper systems of rotation. According to the nitrogen-climate functions discussed in Chap. VI, a permanent augmentation of soil nitrogen and organic matter is most easily accomplished in the North. In the South the high temperatures militate against substantial accumulations of organic matter (29).

The Nature of the Time Function of Soil Fertility.—The characteristic feature of the time curves for soil fertility, *e.g.*, for nitrogen and organic-matter content, is the absence of straight-line relationships or direct proportionalities.

TABLE 68.—EFFECT OF SYSTEMS OF SOIL MANAGEMENT ON THE NITROGEN LEVEL OF THE SOIL
(Sanborn Field, 40 years of cultivation)

Plot number	Cropping system	Pounds of nitrogen per acre (0–8 in.)
	High nitrogen levels	
14	6-year rotation, 6 tons of manure	3,956
34	4-year rotation, 6 tons of manure	3,772
	Medium nitrogen levels	
39	4-year rotation, no treatment	2,875
27	6-year rotation, no treatment	2,691
	Low nitrogen levels	
9	Continuous wheat, no treatment	2,254
17	Continuous corn, no treatment	1,840

Salter and Green describe their previously mentioned curves by means of the equation

$$N = N_0(1 - x)^t \qquad (41)$$

N represents the nitrogen content of the soil at the time t, N_0, is the initial nitrogen content, and x is the annual loss in per cent; t denotes the time in years. For corn, $x = 2.97$ per cent, so that Eq. (41) takes the form

$$N = N_0\left(1 - \frac{2.97}{100}\right)^t = N_0(0.9707)^t \qquad (42)$$

The calculated values of x for the cropping systems studied by Salter and Green are assembled in Table 69. Salter (53) has designated the values of x as soil-productivity indexes and proposed them as a basis for payments on contracted acres under the Agricultural Adjustment Administration. In Ohio, they have been employed throughout the state in county agricultural planning studies.

Salter's equation is the compound-interest law and rests on the assumption that the loss of nitrogen or organic matter is proportional to the amount present. This assumption when expressed in the form of a differential equation becomes

$$-\frac{dN}{dt} = kN; \qquad N = N_0 e^{-kt} \tag{43}$$

This is, indeed, the first assumption one would venture to make in explaining the curves, particularly those observed on the Wooster silt loam in Ohio.

The most significant feature of the above exponential function is the absence of a relative minimum. According to the equation, soil nitrogen declines continuously until it reaches the absolute minimum, which is zero. The plots planted with continuous corn had 2,176 lb. of nitrogen per acre (from 0 to 6⅔ in. deep) in 1894. In 1925, the value was only 840 lb., and, in 1994, or 100 years from the beginning, the soil will contain only approximately 100 lb. per acre according to Eq. (43). Salter's declining curve,

TABLE 69.—PERCENTAGE ANNUAL LOSSES (−) AND GAINS (+) OF SOIL NITROGEN AND ORGANIC MATTER
[Values of $-x$ in Eq. (41)]

Method of cropping	Organic carbon	Nitrogen
Continuous corn	−3.12	−2.97
Continuous wheat	−1.44	−1.56
Continuous oats	−1.41	−1.45
5-year rotation	−0.85	−1.06
3-year rotation	−0.60	−0.69
Hay in 5-year rotation	+1.36	+0.64
Hay in 3-year rotation	+3.25	+2.87

extrapolated to 100 years, is shown in Fig. 124. Of course, we may not be justified in extrapolating Salter's equations, since it was designed for the description of a 30-year period only, but it provides an interesting starting point for long-range considerations.

The study of Burd and Martin's tank experiments (Fig. 113) and the history of agricultural practices in Europe admits of the possibility of a near or quasi equilibrium of nitrogen economy in soils. That is, nitrogen may reach a definite level or stationary state that depends on the productivity variables, climate, plant, soil, and management. The average position of the equilibrium level probably changes with time, but to such a slight extent that the variations may be neglected for agricultural purposes.

On the basis of the equilibrium concept, one may, as a first approximation, assume that nitrogen is lost according to Eq. (43)

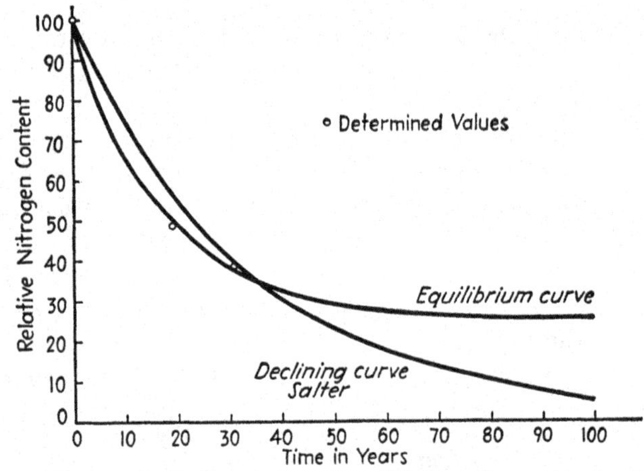

FIG. 124.—This graph depicts the quasi-equilibrium viewpoint of soil fertility as contrasted with the concept of continuous decline.

while at the same time a constant amount of nitrogen is added by fixation from the air. In mathematical language,

$$\frac{dN}{dt} = -k_1N + k_2 \qquad \text{or} \qquad N = N_E - (N_E - N_0)e^{-k_1t} \qquad (44)$$

N_0 is the initial nitrogen content of the soil, t the time, and N_E the equilibrium content, which is equal to k_2/k_1. Salter's equation is a special case of Eq. (44) and is obtained by assigning to k_2 the value of zero ($N_E = 0$). The curve of Eq. (44), based on Salter's data, is shown in Fig. 124. It fits the experimental values satisfactorily. According to the equilibrium concept, the annual loss of soil nitrogen under continuous corn is even greater than Salter's value of 2.97 per cent, namely 5.90 per cent ($k_1 = 0.0608$). However, this magnitude is masked by the fixation process which, in this system of cropping, amounts to 34 lb. of nitrogen per acre per year ($k_2 = 1.55$) as calculated from Fig. 124. More complex equations may have to be devised for other systems of cropping and different combinations of soil-forming factors.

Owing to the approximate proportionality between soil nitrogen and corn yields, it is possible that the form of Eq. (44) may also be used to express functions between productivity and time.

In order to portray in a schematic manner the influence of man on the fertility of soils, Fig. 125 should be studied. It shows hypothetically the accumulation of total nitrogen as a function of time of soil formation under conditions of constant climate, biological environment, and topography. Nature's quasi equi-

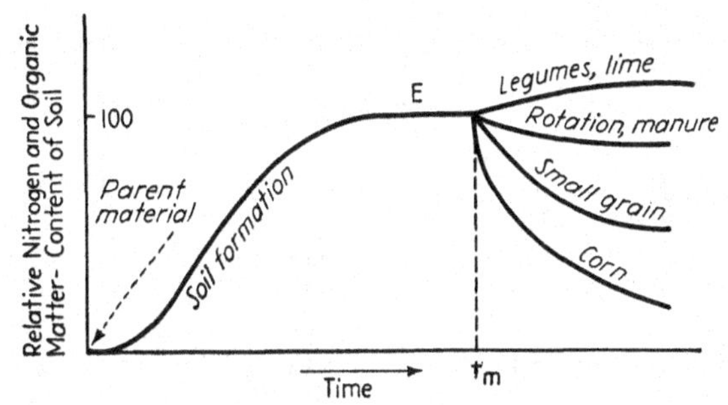

FIG. 125.—Hypothetical illustration of nature's building up of soil fertility and its modifications by man.

librium state is reached at E and supposedly persists for an extensive period of time. At t_m man enters the picture and, by means of various methods of cropping and soil management, changes the nitrogen level of the soil. Continuous cropping of corn has a destructive effect on soil fertility, whereas the intensive use of legumes will preserve the original nitrogen content or may even increase it. Studies of the type illustrated in Fig. 125 are at present in the limelight of agricultural discussions because of the widespread realization of the necessity for conservation of our national resources.

Literature Cited

1. ABBOTT, O. D.: *Sci. Supplement* 91, 2360, p. 10, 1940.
2. ALWAY, F. J., and MCMILLER, P. R.: Interrelationships of soils and forest cover on Star Island, Minnesota, *Soil Sci.*, **36**: 281–295, 387–398, 1933.
3. BARNETTE, R. M., and HESTER, J. B.: Effect of burning upon the accumulation of organic matter in forest soils, *Soil Sci.*, **29**: 281–284, 1930.
4. BAUER, F. C.: Response of Illinois soils to systems of soil treatment, *Illinois Agr. Expt. Sta., Bull.* 362, Urbana, Ill., 1930.

5. Bennett, H. H.: "Soil Conservation," McGraw-Hill Book Company, Inc., New York, 1939.
6. Blöchliger, G.: Mikrobiologische Untersuchungen an verwitternden Schrattenkalkfelsen, Thesis, E. T. H., Zurich, Switzerland, 1931.
7. Bodman, G. B.: The forest floor developed under conditions of summer rainfall deficiency in a Californian pine and fir forest, *Am. Soil Survey Assoc., Bull.* 16: 97–101, 1935.
8. Borden, R. J.: Cane growth studies, *Hawaiian Planters' Record,* **40:** 143–156, 1936.
9. Braun-Blanquet, J.: "Plant Sociology," McGraw-Hill Book Company, Inc., New York, 1932.
10. Bray, R. H.: Unpublished data.
11. Burd, J. S., and Martin, J. C.: Secular and seasonal changes in soils, *Hilgardia,* **5:** 455–509, 1931.
12. Büsgen, M.: "The structure and life of forest trees," John Wiley & Sons, Inc., New York, 1929.
13. Cain, S. A.: Studies on virgin hardwood forest, II, *Am. Midland Naturalist,* **15:** 529–566, 1934.
14. Cain, S. A.: The climax and its complexities, *Am. Midland Naturalist,* **21:** 146–181, 1939.
15. Chandler, R. F.: The calcium content of the foliage of forest trees, *Cornell University, Agr. Expt. Sta. Memoir* 228, Ithaca, N. Y., 1939.
16. Clements, F. E., and Weaver, J. E.: "Experimental Vegetation," Carnegie Institution, 1924.
17. Funchess, M. J.: Alabama Agr. Expt. Sta., Forty-first Annual *Report,* Auburn, Alabama, 1930.
18. Funk, S.: "Die Waldsteppenlandschaften, ihr Wesen und ihre Verbreitung," Verlag D. Friedrichsen, Hamburg, 1927.
19. Gleason, H. A.: The vegetational history of the Middle West, *Ann. Assoc. Am. Geog.,* **12:** 39–85, 1922.
20. Griffith, B. G., Hartwell, E. W., and Shaw, T. E.: The evolution of soils as affected by the old field white pine-mixed hardwood succession in central New England, *Harvard Forest Bull.* 15, 1930.
21. Hardon, H. F.: Factoren, die het organische stof-en het stikstofgehalts van tropische gronden beheerschen, *Medeleelingen alg. Proefst. Landbouw.,* 18, Buitenzorg, 1936.
22. Hesselman, H.: Studier över barrskogens humustäcke, dess egenskaper och beroende av skogvården, *Medd. Statens Skogsförsöcksanstalt,* **22**:169–552, 1925.
23. Hilgard, E. W.: "Soils," The Macmillan Company, New York, 1914.
24. Hoagland, D. R., and Martin, J. C.: Absorption of potassium by plants and fixation by the soil in relation to certain methods for estimating available nutrients, *Trans. Third Intern. Congr. Soil Sci.,* I, 99–103, Oxford, 1935.
25. Hosking, J. S.: The carbon-nitrogen ratios of Australian soils, *Soil Research,* **4:** 253–268, 1935.
26. Hulbert, A. B.: "Soil, its Influence on the History of the United States," Yale University Press, New Haven, 1930.

27. HUNTINGTON, E., WILLIAMS, F. E., and VALKENBURG, S. VAN: "Economic and Social Geography," John Wiley & Sons, Inc., New York, 1933.
28. JENNY, H.: Die Alpinen Böden. In BRAUN-BLANQUET, J. unter Mitwirkung von JENNY, H.: Vegetationsentwicklung und Bodenbildung in der alpinen Stufe der Zentralalpen, *Denkschriften der Schweiz. Naturf. Ges.*, 63 (2):295–340.
29. JENNY, H.: A study on the influence of climate upon the nitrogen and organic matter content of the soil, *Missouri Agr. Expt. Sta. Research Bull.* 152, 1930.
30. JENNY, H.: Soil fertility losses under Missouri conditions, *Missouri Agr. Expt. Sta. Bull.* 324, 1933.
31. JENNY, H., and COWAN, E. W.: The utilization of adsorbed ions by plants, *Science*, **77**: 394–396, 1933.
32. JOFFE, J. S.: "Pedology," Rutgers University Press, New Brunswick, N. J., 1936.
33. KAPPEN, H.: "Bodenazidität," Verlag Julius Springer, Berlin, 1929.
34. KELLEY, W. P.: The reclamation of alkali soils, *California Agr. Expt. Sta. Bull.* 617, 1937.
35. KRAMER, J., and WEAVER, J. E.: Relative efficiency of roots and tops of plants in protecting the soil from erosion, Conservation and Survey Division, University of Nebraska, *Bull.* 12, 1936.
36. LECLERC, J. A., and YODER, P. A.: Tri-local soil exchange experiments with wheat, *Eighth Intern. Congr. Appl. Chem., Orig. Commun.*, **26**: 137–150, 1912.
37. LIPMAN, C. B., and WAYNICK, D. D.: A detailed study of the effects of climate on important properties of soils, *Soil Sci.*, **1**: 5–48, 1916.
38. LIPMAN, J. G., and BLAIR, A. W.: Nitrogen losses under intensive cropping, *Soil Sci.*, **12**:1–19, 1921.
39. LIVINGSTON, E. B., and SHREVE, F.: "The Distribution of Vegetation in the United States as Related to Climatic Conditions," Carnegie Institution, Washington, D. C., No. 284, 1921.
40. LUNT, H. A.: Profile characteristics of New England, Forest Soils. *Connecticut Agr. Expt. Sta. Bull.* 342, 1932.
41. MARBUT, C. F.: Relation of soil type to environment, *Proc. Second Intern. Congr. of Soil Sci.*, **5**:1–6, Moscow, 1932.
42. MARBUT, C. F.: Soils of the United States, *Atlas of American Agriculture*, Part III, Washington, D. C., 1935.
43. MARETT, J. R. DE LA: "Race, Sex and Environment," Hutchinson & Co., Ltd., London, 1936.
44. MILLER, M. F., and KRUSEKOPF, H. H.: The influence of systems of cropping and methods of culture on surface runoff and soil erosion, *Missouri Agr. Expt. Sta. Research Bull.* 177, 1932.
45. MITSCHERLICH, E. A.: Das Wirkungsgesetz der Wachstumsfaktoren, *Landw. Jahrb.*, **56**: 71–92, 1921–22.
46. NIKIFOROFF, C. C.: Weathering and soil formation, *Trans. Third Intern. Congr. Soil Sci.*, **1**: 324–326, Oxford, 1935.
47. RAMANN, E.: Der Boden und sein geographischen Wert, *Mitteilungen der Geographischen Gesellschaft München*, **13**:1–14, 1918–19.

48. Raunkiaer, C.: "The Life Forms of Plants and Statistical Plant Geography," Oxford, 1934.
49. Robinson, G. W.: Soils of Great Britain, *Trans. Third Intern. Congr. of Soil Sci.*, **2**:11–23, London, 1935.
50. Rost, C. O.: Parallelism of the soils developed on the gray drifts of Minnesota, Dissertation, University of Minnesota, 1918.
51. Rost, C. O., and Alway, F. J.: Minnesota glacial soil studies, I, *Soil Sci.*, **11**:161–200, 1921.
52. Salmon, S. C.: Corn production in Kansas, *Kansas Agr. Expt. Sta. Bull.* 238, 1926.
53. Salter, R. M.: Our heritage, *Ohio Agr. Expt. Service Bull.* 175, 1936.
54. Salter, R. M., and Green, T. C.: Factors affecting the accumulation and loss of nitrogen and organic carbon in cropped soils, *J. Am. Soc. Agron.*, **25**: 622–630, 1933.
55. Scholz, H. F.: Physical properties of the cover soils on the Black Rock Forest, *Black Rock Forest Bull.* 2, Cornwall-on-the-Hudson, N. Y., 1931.
56. Shantz, H. L.: The natural vegetation of the Great Plains region, *Ann. Assoc. Am. Geog.*, **8**: 81–107, 1923.
57. Shantz, H. L., and Zon, R.: Natural Vegetation, *Atlas of American Agriculture*, Part I, Washington, D. C., 1924.
58. Sherlock, R. L.: "Man's Influence on the Earth," T. Thornton Butterworth, Ltd., London, 1931.
59. Stepanov, N. N.: Die Zersetzung des Laubabfalles der Bäume und Sträucher, *Proc. Second Intern. Congr. Soil Sci.*, **5**: 300–305, 1932.
60. Swederski, W.: Untersuchungen über die Gebirgsböden in den Ostkarpaten, I, *Mémoires de l'Institut National Polonais d'Economie Rurale a Pulawy*, **12**:I, 115–154, 1931.
61. Tamm, O.: Der braune Waldboden in Schweden, *Proc. Second Intern. Congr. Soil Sci.*, **5**:178–189, Moscow, 1932.
62. Transeau, E. N.: Forest centers of eastern America, *Am. Naturalist*, **39**: 875–889, 1905.
63. Vageler, P.: "Grundriss der tropischen und subtropischen Bodenkunde," Verlagsgesellschaft für Ackerbau, Berlin, 1930.
64. Vries, O. de: Ertragskurven und Ertragsgesetze, *Trans. Fourth Comm. Intern. Soc. Soil Sci.*, 11–28, Stockholm, 1939.
65. Weaver, J. E., and Clements, F. E.: "Plant Ecology," McGraw-Hill Book Company, Inc., New York, 1938.
66. Weaver, J. E., Hougen, V. H., and Weldon, M. D.: Relation of root distribution to organic matter in prairie soil, *Botan. Gaz.*, **96**: 389–420, 1935.
67. Weber, R.: Die Bedeutung des Waldes und die Aufgaben der Forstwirtschaft, *Handbuch der Forstwissenschaft*, Vol. 1, Tübingen, 1913.
68. Willcox, O. W.: "A B C of Agrobiology," W. W. Norton & Company, Inc., New York, 1937.
69. Zederbauer, E.: Ein Beitrag zur Kenntnis des Wurzelwachstums der Fichte, *Centr. ges. Forstwesen*, **46**: 336–337, 1920.

CHAPTER VIII

CONCLUSIONS

In this chapter, we propose to touch upon certain aspects of wider scope than hitherto discussed and see what general inferences can be drawn from the material presented in the preceding pages.

Retrospect.—Since the turn of the last century, the concept of soil-forming factors has been accorded a prominent position in pedological literature. In this book, an attempt has been made to develop the idea of soil formers into a useful tool of research. However, the prevailing vagueness regarding the definitions of soil-forming factors prevented quantitative application, and it was found necessary to develop at the outset a more rigorous terminology. Accordingly, in the introductory chapter, the soil formers were given the status of independent variables that define the state of a soil system. In this interpretation, the common soil properties become dependent variables and may be expressed as functions of soil-forming factors. In this manner, it is possible to obtain quantitative correlations between soil properties and soil-forming factors. The major portion of this book is devoted to the discussion of such relationships. The results clearly prove the usefulness of this method of soil research.

Restrictions of Applicability.—The requirements necessary for the establishment of such correlations can be rigidly fulfilled only under controlled experimental conditions. In the field, we must be satisfied with approximations and general trends. Not all parts of the world are equally well suited for the evaluation of quantitative functions between soil properties and soil-forming factors. Nature does not always offer the soil formers in suitable combinations for solving the fundamental equation. Probably the most trustworthy functions on hand are those pertaining to time relationships because of the certainty of the qualitative and quantitative nature of the soil-forming factors involved. A much less degree of reliability is inherent in the soil property-climate functions because of the difficulty of estimating the variables, parent material and time. If arrangements could be made to expose given parent materials to the various climates of the earth and to study the changes of soil properties with time, if only for a century or two, a great number of important soil-climate functions could be derived with high accuracy. This may well be the approach of the future.

Freedom from Theories.—Functional analysis of soils is not based on physical, chemical, or biological theories. The entire analysis involves but one hypothesis, namely, the assumption that the variables climate, organisms, topography, parent material, and time, plus some accessory factors such as seasonal variations of climate and ground-water table, suffice to define any soil. The functions themselves are purely observational, and their validity depends solely on the skill and experience of the investigators in selecting suitable areas for study. This represents a distinct advantage over other systems of soil descriptions that rely on speculative mechanisms of soil-forming processes.

Soil Maps versus Soil Functions.—The goal of the soil geographer is the assemblage of soil knowledge in the form of a map. In contrast, the goal of the "functionalist" is the assemblage of soil knowledge in the form of a curve or an equation. These objectives are clearly brought out in a comparison between Marbut's Soils of the United States and the present treatise on soil-forming factors. The former abounds with carefully executed maps of various scales, whereas the latter displays a sequence of graphs and equations. Both methods of approach have their merits, and nothing would be gained by playing one system against the other. Soil maps display the areal arrangement of soil properties and types but give no insight into "causal" relationships. The curve, on the other hand, reveals the dependency of soil properties on soil-forming factors, but the conversion of such fundamental knowledge to specific field conditions is impossible unless the areal distribution of the soil formers is known. Clearly, it is the union of the geographic and the functional method that provides the most effective means of pedological research.

Relation to Other Sciences.—The approach outlined in this book stresses the intimate connections between soil science and related sciences. Although pedology is greatly indebted to climatology, geology, botany, etc., it has, in return, stimulated the growth of the parental sciences and is likely to continue to do so with increasing vigor.

It is not surprising that this influence has been especially marked in the case of plant ecology, which is concerned, to a large extent, with the dependent features of vegetation. The recognition of the parallelism of plant successions and soil development has given new impetus to the study of the distribution of plants, and the idea of soil equilibrium has strengthened the concept of the plant climax. As regards climatology the discovery of the climatic element in soil formation has fostered the search for quantitative climatic indexes and has intensified the desire for more satisfactory measurements of the magnitudes of evaporation. At present, the evaluations of microclimate and of soil climate are gaining momentum and hold considerable promise for the future.

Pedology is likely to fertilize various branches of the geological sciences. Sedimentary petrology already has benefited from the investigations of the nature and of the formation of colloidal clays. The explanations of the origin of geologic strata, such as laterite and ferruginous deposits in general, must be formulated in conjunction with pedological research. Geomorphology, particularly its aspects relating to erosion and denudation, is paying increasing attention to the physical and chemical properties of the soil.

Great interest and vast importance are commanded by the position of pedology in relation to human enterprises in general and to the social sciences in particular. Aside from the role of soil science in the pursuit of the technical phases of agriculture, the significance of soils in influencing social structures, settlement policies, economic questions, etc., is receiving increased recognition, a trend that finds concrete manifestation in the participation of numerous pedologists in national and state planning commissions.

Soil-forming Factors in Relation to Soil Classification.—Glinka, Ramann, Marbut, Shaw, de Sigmond and many other pedologists have devoted a considerable portion of their time to the formulation of systems of soil classification. This interest in the classification of soils is justified by the enormous practical importance of workable systems and by the intellectual satisfaction they give to the human mind, which has an inherent desire for orderliness and systematization. Although new classifications are continually proposed, surprisingly little has been done regarding the formulation of principles by which one might judge and compare the intrinsic value of the various systems proposed. It is believed that functional analysis throws some light on this complicated but nonetheless important phase of theoretical soil science.

A survey of systems of soil classification discloses two major principles of arrangements. One is based on soil properties (*s* values); the other correlates soils in terms of soil-forming factors. In other words, there are classifications that are restricted to dependent variables and others that rest essentially on the independent variables *cl*, *o*, *r*, *p*, and *t*. A number of systems attempt combinations of these two extreme possibilities.

Systems Based on Soil Properties (*s values*).—Soil differentiation is sometimes based on a single soil property such as color or texture. More common are combination systems that, in addition to color and texture, include organic matter, pH, exchangeable cations, silica-alumina ratio, etc. Since the number of soil properties of the *s* type is very large, the task of classifying soils crystallizes around the question: Which of the many *s* values should be selected to form a system of classification?

As far as practical systems are concerned the answer is obvious. A soil-fertility specialist would probably select available potassium, phosphorus, nitrates, etc. An irrigation engineer would focus attention on factors relating to water permeability and water retention; a road builder would classify soils according to clay content, plasticity, and properties connected with swelling and shrinkage. In other words, the choice of soil properties is determined by practical considerations.

A different principle is chosen in the scientific systems of soil classification. In these, the selection of soil characteristics rests on theories of weathering reactions and processes of soil formation. The resulting arrangements of soils are known as "genetic systems" of soil classification.

The recent scheme of de Sigmond may be taken as a specific illustration. De Sigmond's selection of *s* properties is guided by chemical processes of humification and mineralization, as illustrated by the two following sequences:

a. Raw organic matter → humified matter → CO_2,

b. Raw mineral material → siallites → allites.

The main groups and the subgroups of de Sigmond's system are summarized in Table 70. It is important to note that the "main groups" rely on soil properties only. No reference is made to hypothetical degrees of maturity or to features of zonality.

TABLE 70.—DE SIGMOND'S SYSTEM OF SOIL CLASSIFICATION (Abbreviated)

Main Groups	Subgroups
I. *Organic soils*, the organic matter (C × 1.72) exceeding 25–30 per cent, and the depth of the organic layer 25–30 cm.	1. *Raw organic soils*, characterized by visible botanical structure of the surface layer
	2. *Humified organic soils*, the botanical structure of the surface layer being invisible
II. *Soils of mixed composition*, the organic matter not exceeding 25–30 per cent, and the depth of the surface layer, if organic itself, not exceeding 25–30 cm.	3. *Raw soils of mixed origin*, in which decomposition of the raw materials is almost absent or the soil has no characteristic profile
	4. *Humo-siallitic* soils characterized by various degrees of humification, and in the HCl extract $SiO_2:Al_2O_3 > 2:1$
	5. *Ferric-siallitic soils*, the typical color of the soils being due to Fe_2O_3 compounds, and $SiO_2:Al_2O_3 > 2:1$ in the HCl extract
	6. *Allitic soils*, characterized by extreme soil leaching and $SiO_2:Al_2O_3 < 2:1$ in the HCl-extract
III. *Purely mineral soils*, containing no living or dead organic matter	7. *Raw, purely mineral soils*
	8. *Purely mineral soils with initial weathering*
	9. Purely mineral soils incrusted with some end products (salts) of mineral weathering

Preceding de Sigmond, Gedroiz proposed a system of soil classification that placed the main emphasis on the chemical composition of the colloidal particles of the soil. His main groups were as follows:

I. Soils saturated with bases.

a. Ca-Mg-saturated (*e.g.*, chernozems),

b. Na-saturated (saline soils, alkali soils, solonetz).

II. Soils unsaturated with bases.
 a. Wide silica-alumina ratio (podsols),
 b. Narrow silica-alumina ratio (lateritic soils, yellow and red soils).

Although Gedroiz's grouping of soils appears theoretically sound, it lacks a wider appeal, because it excludes visible soil properties. Soils cannot be identified in the field. Unless samples are sent to a laboratory and subjected to chemical analysis, classification cannot be attained.

Generally speaking, the value of the genetic systems of soil classification based on *s* properties is governed by the validity of the theoretical assumptions that guide the selection of criteria. Depending on the taxonomist's knowledge of soil-forming processes, the systems will be found satisfactory or wanting in completeness. It appears to the author that the universal classification of soils based on strictly scientific principles must await fulfillment until more precise knowledge of the physical, chemical, and biological reactions occurring in soils is at hand.

Systems Based on Soil-forming Factors.—Groupings such as arid and humid soils, arctic and tropical soils, loessal and limestone soils, prairie soils and forest soils do not refer to soil characteristics (*s* values) but to climate, parent material, and vegetation, in short, to soil-forming factors. Many soil scientists violently oppose the use of soil-forming factors as criteria of soil classification, branding the efforts as unscientific and illogical. They contend that soils, like plants and animals, should be classified according to their own properties (*s* values) and nothing else. In reply, the author contends that soils are not living systems. They have no power of reproduction and no heredity. Unlike organisms, their properties are solely determined by the soil-forming factors. In view of the numerous correlations between soil properties and soil-forming factors presented in the preceding chapters, a wholly negative attitude toward the use of soil formers in soil classification seems not justified on theoretical grounds. As a matter of fact, the majority of the popular systems of soil classification, especially in the United States, involve the concept of soil-forming factors. In previous chapters, we have had several opportunities to describe classification systems based on soil-forming factors and to point out their advantages and limitations.

In recent years, a number of attempts have been made to evolve schemes of classification that are more closely in line with the principles of functional analysis. In these groupings, the individual soils are arranged primarily in terms of a single soil-forming factor that appears as a variable. Examples of this type are presented by Shaw's classification of soils into families (see page 49) and probably Milne's grouping of soils as catenas (see page 98). The family, according to Shaw, comprises soils that possess similar soil-forming factors save one, namely, time or degree of maturity. Milne's catena represents an association of soil types that are derived from similar parent materials but differ in accordance with the factor topography or, more specifically, slope and drainage conditions.

At the present state of soil knowledge, it is difficult to predict the future trend of scientific systems of soil classification. The schemes stressing soil properties will have to keep pace with the advancement of theories of soil formation, whereas the groupings according to soil-forming factors are handicapped by the exact determination of the initial (parent material) and final (mature soil) stages of soil development.

Restrictions Placed on Functional Relationships among s Properties.—In all parts of the world, efforts have been made to discover quantitative relationships among soil properties of the *s* type. Some of the most widely studied functions involve the following pairs of properties: total nitrogen versus total organic carbon (known as carbon-nitrogen ratio), base exchange capacity versus silica-alumina ratio of clay particles, pH of soil versus lime requirement, crop yields versus quick tests, erosiveness versus dispersion ratio, etc.

These relationships deal exclusively with *s* properties, in other words, with dependent variables. Although it is generally admitted that none of these functions has proved entirely satisfactory, apparently no one has questioned the soundness of the approach. Interestingly enough, the soil-forming factor Eq. (4) sheds new light on this problem and permits enunciation of the following postulate: *It is improbable that a function between two s properties* that possesses general validity will ever be found.* To

* Provided that the value of one property does not uniquely determine the value of another, such as the percentage of organic and of inorganic matter, or the content of positive and of negative valences.

cite a specific example: If one were to collect at random a large number of soils from all parts of the world and analyze for two s properties, such as carbon and nitrogen, no correlation between these two properties would be found. The detailed proof of this theorem requires the use of Jacobians and therefore will not be attempted here. It will suffice to formulate briefly the essential steps in the trend of thought. Given two soil properties s_m and s_n, both of which obey Eq. (4)

$$s_m = f_m(cl, o, r, p, t)$$

and

$$s_n = f_n(cl, o, r, p, t)$$

We ask the question: Does any continuous relationship between s_m and s_n exist? Is

$$s_m = f(s_n)?$$

The answer is no, because it is improbable that the Jacobians vanish. An illustration may be found in Chap. VI in which the nitrogen and the clay content of soils are expressed as functions of temperature and moisture. The Jacobian for the four variables nitrogen, clay, temperature, and moisture does not vanish, and no general function between nitrogen content and clay content exists. Evidently, if the theorem should be universally true, it would cast serious doubt on the general validity of certain types of physical, chemical, and biological soil investigations.

The absence of general relationships among s properties does not exclude the existence of limited correlations. These may be found in areas within which some of the soil-forming factors are relatively constant. Suppose we restrain all soil-forming factors except one that is left to vary. Selecting t for the sake of illustration, we may write

$$s_m = f_\mu(t)_{cl_1, o_1, r_1, p_1}$$

and

$$s_n = f_\nu(t)_{cl_1, o_1, r_1, p_1}$$

Under these circumstances a continuous function between s_m and s_n is likely to obtain. Searches for relationships among two s properties will be rewarded, provided care is taken that the soil-forming factors are controlled. The greater the number of constant soil-forming factors the greater is the likelihood of finding satisfactory correlations. In the central part of the United States fairly satisfactory correlations between *s* properties frequently have been reported. These functions are valid, because these regions are characterized by a relative constancy of soil formers. In other parts of the country, notably on the Pacific Coast, correlations are notoriously poor. The combination of high mountain massives and a vast ocean produces such wide variations in climate, organisms, parent material, and topography that any effort to establish general functions among soil properties of the *s* type must lead to disappointments.

AUTHOR INDEX

A

B

C

D

E

F

G

H

I

J

K

L

M

SUBJECT INDEX

D

O

P

A CATALOG OF SELECTED

DOVER BOOKS

IN SCIENCE AND MATHEMATICS

Astronomy

CHARIOTS FOR APOLLO: The NASA History of Manned Lunar Spacecraft to 1969, Courtney G. Brooks, James M. Grimwood, and Loyd S. Swenson, Jr. This illustrated history by a trio of experts is the definitive reference on the Apollo spacecraft and lunar modules. It traces the vehicles' design, development, and operation in space. More than 100 photographs and illustrations. 576pp. 6 3/4 x 9 1/4. 0-486-46756-2

EXPLORING THE MOON THROUGH BINOCULARS AND SMALL TELESCOPES, Ernest H. Cherrington, Jr. Informative, profusely illustrated guide to locating and identifying craters, rills, seas, mountains, other lunar features. Newly revised and updated with special section of new photos. Over 100 photos and diagrams. 240pp. 8 1/4 x 11. 0-486-24491-1

WHERE NO MAN HAS GONE BEFORE: A History of NASA's Apollo Lunar Expeditions, William David Compton. Introduction by Paul Dickson. This official NASA history traces behind-the-scenes conflicts and cooperation between scientists and engineers. The first half concerns preparations for the Moon landings, and the second half documents the flights that followed Apollo 11. 1989 edition. 432pp. 7 x 10. 0-486-47888-2

APOLLO EXPEDITIONS TO THE MOON: The NASA History, Edited by Edgar M. Cortright. Official NASA publication marks the 40th anniversary of the first lunar landing and features essays by project participants recalling engineering and administrative challenges. Accessible, jargon-free accounts, highlighted by numerous illustrations. 336pp. 8 3/8 x 10 7/8. 0-486-47175-6

ON MARS: Exploration of the Red Planet, 1958-1978--The NASA History, Edward Clinton Ezell and Linda Neuman Ezell. NASA's official history chronicles the start of our explorations of our planetary neighbor. It recounts cooperation among government, industry, and academia, and it features dozens of photos from Viking cameras. 560pp. 6 3/4 x 9 1/4. 0-486-46757-0

ARISTARCHUS OF SAMOS: The Ancient Copernicus, Sir Thomas Heath. Heath's history of astronomy ranges from Homer and Hesiod to Aristarchus and includes quotes from numerous thinkers, compilers, and scholasticists from Thales and Anaximander through Pythagoras, Plato, Aristotle, and Heraclides. 34 figures. 448pp. 5 3/8 x 8 1/2. 0-486-43886-4

AN INTRODUCTION TO CELESTIAL MECHANICS, Forest Ray Moulton. Classic text still unsurpassed in presentation of fundamental principles. Covers rectilinear motion, central forces, problems of two and three bodies, much more. Includes over 200 problems, some with answers. 437pp. 5 3/8 x 8 1/2. 0-486-64687-4

BEYOND THE ATMOSPHERE: Early Years of Space Science, Homer E. Newell. This exciting survey is the work of a top NASA administrator who chronicles technological advances, the relationship of space science to general science, and the space program's social, political, and economic contexts. 528pp. 6 3/4 x 9 1/4. 0-486-47464-X

STAR LORE: Myths, Legends, and Facts, William Tyler Olcott. Captivating retellings of the origins and histories of ancient star groups include Pegasus, Ursa Major, Pleiades, signs of the zodiac, and other constellations. "Classic." – *Sky & Telescope.* 58 illustrations. 544pp. 5 3/8 x 8 1/2. 0-486-43581-4

A COMPLETE MANUAL OF AMATEUR ASTRONOMY: Tools and Techniques for Astronomical Observations, P. Clay Sherrod with Thomas L. Koed. Concise, highly readable book discusses the selection, set-up, and maintenance of a telescope; amateur studies of the sun; lunar topography and occultations; and more. 124 figures. 26 halftones. 37 tables. 335pp. 6 1/2 x 9 1/4. 0-486-42820-6

Chemistry

MOLECULAR COLLISION THEORY, M. S. Child. This high-level monograph offers an analytical treatment of classical scattering by a central force, quantum scattering by a central force, elastic scattering phase shifts, and semi-classical elastic scattering. 1974 edition. 310pp. 5 3/8 x 8 1/2. 0-486-69437-2

HANDBOOK OF COMPUTATIONAL QUANTUM CHEMISTRY, David B. Cook. This comprehensive text provides upper-level undergraduates and graduate students with an accessible introduction to the implementation of quantum ideas in molecular modeling, exploring practical applications alongside theoretical explanations. 1998 edition. 832pp. 5 3/8 x 8 1/2. 0-486-44307-8

RADIOACTIVE SUBSTANCES, Marie Curie. The celebrated scientist's thesis, which directly preceded her 1903 Nobel Prize, discusses establishing atomic character of radioactivity; extraction from pitchblende of polonium and radium; isolation of pure radium chloride; more. 96pp. 5 3/8 x 8 1/2. 0-486-42550-9

CHEMICAL MAGIC, Leonard A. Ford. Classic guide provides intriguing entertainment while elucidating sound scientific principles, with more than 100 unusual stunts: cold fire, dust explosions, a nylon rope trick, a disappearing beaker, much more. 128pp. 5 3/8 x 8 1/2. 0-486-67628-5

ALCHEMY, E. J. Holmyard. Classic study by noted authority covers 2,000 years of alchemical history: religious, mystical overtones; apparatus; signs, symbols, and secret terms; advent of scientific method, much more. Illustrated. 320pp. 5 3/8 x 8 1/2. 0-486-26298-7

CHEMICAL KINETICS AND REACTION DYNAMICS, Paul L. Houston. This text teaches the principles underlying modern chemical kinetics in a clear, direct fashion, using several examples to enhance basic understanding. Solutions to selected problems. 2001 edition. 352pp. 8 3/8 x 11. 0-486-45334-0

PROBLEMS AND SOLUTIONS IN QUANTUM CHEMISTRY AND PHYSICS, Charles S. Johnson and Lee G. Pedersen. Unusually varied problems, with detailed solutions, cover of quantum mechanics, wave mechanics, angular momentum, molecular spectroscopy, scattering theory, more. 280 problems, plus 139 supplementary exercises. 430pp. 6 1/2 x 9 1/4. 0-486-65236-X

ELEMENTS OF CHEMISTRY, Antoine Lavoisier. Monumental classic by the founder of modern chemistry features first explicit statement of law of conservation of matter in chemical change, and more. Facsimile reprint of original (1790) Kerr translation. 539pp. 5 3/8 x 8 1/2. 0-486-64624-6

MAGNETISM AND TRANSITION METAL COMPLEXES, F. E. Mabbs and D. J. Machin. A detailed view of the calculation methods involved in the magnetic properties of transition metal complexes, this volume offers sufficient background for original work in the field. 1973 edition. 240pp. 5 3/8 x 8 1/2. 0-486-46284-6

GENERAL CHEMISTRY, Linus Pauling. Revised third edition of classic first-year text by Nobel laureate. Atomic and molecular structure, quantum mechanics, statistical mechanics, thermodynamics correlated with descriptive chemistry. Problems. 992pp. 5 3/8 x 8 1/2. 0-486-65622-5

ELECTROLYTE SOLUTIONS: Second Revised Edition, R. A. Robinson and R. H. Stokes. Classic text deals primarily with measurement, interpretation of conductance, chemical potential, and diffusion in electrolyte solutions. Detailed theoretical interpretations, plus extensive tables of thermodynamic and transport properties. 1970 edition. 590pp. 5 3/8 x 8 1/2. 0-486-42225-9

Engineering

FUNDAMENTALS OF ASTRODYNAMICS, Roger R. Bate, Donald D. Mueller, and Jerry E. White. Teaching text developed by U.S. Air Force Academy develops the basic two-body and n-body equations of motion; orbit determination; classical orbital elements, coordinate transformations; differential correction; more. 1971 edition. 455pp. 5 3/8 x 8 1/2. 0-486-60061-0

INTRODUCTION TO CONTINUUM MECHANICS FOR ENGINEERS: Revised Edition, Ray M. Bowen. This self-contained text introduces classical continuum models within a modern framework. Its numerous exercises illustrate the governing principles, linearizations, and other approximations that constitute classical continuum models. 2007 edition. 320pp. 6 1/8 x 9 1/4. 0-486-47460-7

ENGINEERING MECHANICS FOR STRUCTURES, Louis L. Bucciarelli. This text explores the mechanics of solids and statics as well as the strength of materials and elasticity theory. Its many design exercises encourage creative initiative and systems thinking. 2009 edition. 320pp. 6 1/8 x 9 1/4. 0-486-46855-0

FEEDBACK CONTROL THEORY, John C. Doyle, Bruce A. Francis and Allen R. Tannenbaum. This excellent introduction to feedback control system design offers a theoretical approach that captures the essential issues and can be applied to a wide range of practical problems. 1992 edition. 224pp. 6 1/2 x 9 1/4. 0-486-46933-6

THE FORCES OF MATTER, Michael Faraday. These lectures by a famous inventor offer an easy-to-understand introduction to the interactions of the universe's physical forces. Six essays explore gravitation, cohesion, chemical affinity, heat, magnetism, and electricity. 1993 edition. 96pp. 5 3/8 x 8 1/2. 0-486-47482-8

DYNAMICS, Lawrence E. Goodman and William H. Warner. Beginning engineering text introduces calculus of vectors, particle motion, dynamics of particle systems and plane rigid bodies, technical applications in plane motions, and more. Exercises and answers in every chapter. 619pp. 5 3/8 x 8 1/2. 0-486-42006-X

ADAPTIVE FILTERING PREDICTION AND CONTROL, Graham C. Goodwin and Kwai Sang Sin. This unified survey focuses on linear discrete-time systems and explores natural extensions to nonlinear systems. It emphasizes discrete-time systems, summarizing theoretical and practical aspects of a large class of adaptive algorithms. 1984 edition. 560pp. 6 1/2 x 9 1/4. 0-486-46932-8

INDUCTANCE CALCULATIONS, Frederick W. Grover. This authoritative reference enables the design of virtually every type of inductor. It features a single simple formula for each type of inductor, together with tables containing essential numerical factors. 1946 edition. 304pp. 5 3/8 x 8 1/2. 0-486-47440-2

THERMODYNAMICS: Foundations and Applications, Elias P. Gyftopoulos and Gian Paolo Beretta. Designed by two MIT professors, this authoritative text discusses basic concepts and applications in detail, emphasizing generality, definitions, and logical consistency. More than 300 solved problems cover realistic energy systems and processes. 800pp. 6 1/8 x 9 1/4. 0-486-43932-1

THE FINITE ELEMENT METHOD: Linear Static and Dynamic Finite Element Analysis, Thomas J. R. Hughes. Text for students without in-depth mathematical training, this text includes a comprehensive presentation and analysis of algorithms of time-dependent phenomena plus beam, plate, and shell theories. Solution guide available upon request. 672pp. 6 1/2 x 9 1/4. 0-486-41181-8

Mathematics-Bestsellers

HANDBOOK OF MATHEMATICAL FUNCTIONS: with Formulas, Graphs, and Mathematical Tables, Edited by Milton Abramowitz and Irene A. Stegun. A classic resource for working with special functions, standard trig, and exponential logarithmic definitions and extensions, it features 29 sets of tables, some to as high as 20 places. 1046pp. 8 x 10 1/2. 0-486-61272-4

ABSTRACT AND CONCRETE CATEGORIES: The Joy of Cats, Jiri Adamek, Horst Herrlich, and George E. Strecker. This up-to-date introductory treatment employs category theory to explore the theory of structures. Its unique approach stresses concrete categories and presents a systematic view of factorization structures. Numerous examples. 1990 edition, updated 2004. 528pp. 6 1/8 x 9 1/4. 0-486-46934-4

MATHEMATICS: Its Content, Methods and Meaning, A. D. Aleksandrov, A. N. Kolmogorov, and M. A. Lavrent'ev. Major survey offers comprehensive, coherent discussions of analytic geometry, algebra, differential equations, calculus of variations, functions of a complex variable, prime numbers, linear and non-Euclidean geometry, topology, functional analysis, more. 1963 edition. 1120pp. 5 3/8 x 8 1/2. 0-486-40916-3

INTRODUCTION TO VECTORS AND TENSORS: Second Edition--Two Volumes Bound as One, Ray M. Bowen and C.-C. Wang. Convenient single-volume compilation of two texts offers both introduction and in-depth survey. Geared toward engineering and science students rather than mathematicians, it focuses on physics and engineering applications. 1976 edition. 560pp. 6 1/2 x 9 1/4. 0-486-46914-X

AN INTRODUCTION TO ORTHOGONAL POLYNOMIALS, Theodore S. Chihara. Concise introduction covers general elementary theory, including the representation theorem and distribution functions, continued fractions and chain sequences, the recurrence formula, special functions, and some specific systems. 1978 edition. 272pp. 5 3/8 x 8 1/2. 0-486-47929-3

ADVANCED MATHEMATICS FOR ENGINEERS AND SCIENTISTS, Paul DuChateau. This primary text and supplemental reference focuses on linear algebra, calculus, and ordinary differential equations. Additional topics include partial differential equations and approximation methods. Includes solved problems. 1992 edition. 400pp. 7 1/2 x 9 1/4. 0-486-47930-7

PARTIAL DIFFERENTIAL EQUATIONS FOR SCIENTISTS AND ENGINEERS, Stanley J. Farlow. Practical text shows how to formulate and solve partial differential equations. Coverage of diffusion-type problems, hyperbolic-type problems, elliptic-type problems, numerical and approximate methods. Solution guide available upon request. 1982 edition. 414pp. 6 1/8 x 9 1/4. 0-486-67620-X

VARIATIONAL PRINCIPLES AND FREE-BOUNDARY PROBLEMS, Avner Friedman. Advanced graduate-level text examines variational methods in partial differential equations and illustrates their applications to free-boundary problems. Features detailed statements of standard theory of elliptic and parabolic operators. 1982 edition. 720pp. 6 1/8 x 9 1/4. 0-486-47853-X

LINEAR ANALYSIS AND REPRESENTATION THEORY, Steven A. Gaal. Unified treatment covers topics from the theory of operators and operator algebras on Hilbert spaces; integration and representation theory for topological groups; and the theory of Lie algebras, Lie groups, and transform groups. 1973 edition. 704pp. 6 1/8 x 9 1/4. 0-486-47851-3

A SURVEY OF INDUSTRIAL MATHEMATICS, Charles R. MacCluer. Students learn how to solve problems they'll encounter in their professional lives with this concise single-volume treatment. It employs MATLAB and other strategies to explore typical industrial problems. 2000 edition. 384pp. 5 3/8 x 8 1/2. 0-486-47702-9

NUMBER SYSTEMS AND THE FOUNDATIONS OF ANALYSIS, Elliott Mendelson. Geared toward undergraduate and beginning graduate students, this study explores natural numbers, integers, rational numbers, real numbers, and complex numbers. Numerous exercises and appendixes supplement the text. 1973 edition. 368pp. 5 3/8 x 8 1/2. 0-486-45792-3

A FIRST LOOK AT NUMERICAL FUNCTIONAL ANALYSIS, W. W. Sawyer. Text by renowned educator shows how problems in numerical analysis lead to concepts of functional analysis. Topics include Banach and Hilbert spaces, contraction mappings, convergence, differentiation and integration, and Euclidean space. 1978 edition. 208pp. 5 3/8 x 8 1/2. 0-486-47882-3

FRACTALS, CHAOS, POWER LAWS: Minutes from an Infinite Paradise, Manfred Schroeder. A fascinating exploration of the connections between chaos theory, physics, biology, and mathematics, this book abounds in award-winning computer graphics, optical illusions, and games that clarify memorable insights into self-similarity. 1992 edition. 448pp. 6 1/8 x 9 1/4. 0-486-47204-3

SET THEORY AND THE CONTINUUM PROBLEM, Raymond M. Smullyan and Melvin Fitting. A lucid, elegant, and complete survey of set theory, this three-part treatment explores axiomatic set theory, the consistency of the continuum hypothesis, and forcing and independence results. 1996 edition. 336pp. 6 x 9. 0-486-47484-4

DYNAMICAL SYSTEMS, Shlomo Sternberg. A pioneer in the field of dynamical systems discusses one-dimensional dynamics, differential equations, random walks, iterated function systems, symbolic dynamics, and Markov chains. Supplementary materials include PowerPoint slides and MATLAB exercises. 2010 edition. 272pp. 6 1/8 x 9 1/4. 0-486-47705-3

ORDINARY DIFFERENTIAL EQUATIONS, Morris Tenenbaum and Harry Pollard. Skillfully organized introductory text examines origin of differential equations, then defines basic terms and outlines general solution of a differential equation. Explores integrating factors; dilution and accretion problems; Laplace Transforms; Newton's Interpolation Formulas, more. 818pp. 5 3/8 x 8 1/2. 0-486-64940-7

MATROID THEORY, D. J. A. Welsh. Text by a noted expert describes standard examples and investigation results, using elementary proofs to develop basic matroid properties before advancing to a more sophisticated treatment. Includes numerous exercises. 1976 edition. 448pp. 5 3/8 x 8 1/2. 0-486-47439-9

THE CONCEPT OF A RIEMANN SURFACE, Hermann Weyl. This classic on the general history of functions combines function theory and geometry, forming the basis of the modern approach to analysis, geometry, and topology. 1955 edition. 208pp. 5 3/8 x 8 1/2. 0-486-47004-0

THE LAPLACE TRANSFORM, David Vernon Widder. This volume focuses on the Laplace and Stieltjes transforms, offering a highly theoretical treatment. Topics include fundamental formulas, the moment problem, monotonic functions, and Tauberian theorems. 1941 edition. 416pp. 5 3/8 x 8 1/2. 0-486-47755-X

Browse over 9,000 books at www.doverpublications.com

Mathematics–Logic and Problem Solving

PERPLEXING PUZZLES AND TANTALIZING TEASERS, Martin Gardner. Ninety-three riddles, mazes, illusions, tricky questions, word and picture puzzles, and other challenges offer hours of entertainment for youngsters. Filled with rib-tickling drawings. Solutions. 224pp. 5 3/8 x 8 1/2. 0-486-25637-5

MY BEST MATHEMATICAL AND LOGIC PUZZLES, Martin Gardner. The noted expert selects 70 of his favorite "short" puzzles. Includes The Returning Explorer, The Mutilated Chessboard, Scrambled Box Tops, and dozens more. Complete solutions included. 96pp. 5 3/8 x 8 1/2. 0-486-28152-3

THE LADY OR THE TIGER?: and Other Logic Puzzles, Raymond M. Smullyan. Created by a renowned puzzle master, these whimsically themed challenges involve paradoxes about probability, time, and change; metapuzzles; and self-referentiality. Nineteen chapters advance in difficulty from relatively simple to highly complex. 1982 edition. 240pp. 5 3/8 x 8 1/2. 0-486-47027-X

SATAN, CANTOR AND INFINITY: Mind-Boggling Puzzles, Raymond M. Smullyan. A renowned mathematician tells stories of knights and knaves in an entertaining look at the logical precepts behind infinity, probability, time, and change. Requires a strong background in mathematics. Complete solutions. 288pp. 5 3/8 x 8 1/2. 0-486-47036-9

THE RED BOOK OF MATHEMATICAL PROBLEMS, Kenneth S. Williams and Kenneth Hardy. Handy compilation of 100 practice problems, hints and solutions indispensable for students preparing for the William Lowell Putnam and other mathematical competitions. Preface to the First Edition. Sources. 1988 edition. 192pp. 5 3/8 x 8 1/2. 0-486-69415-1

KING ARTHUR IN SEARCH OF HIS DOG AND OTHER CURIOUS PUZZLES, Raymond M. Smullyan. This fanciful, original collection for readers of all ages features arithmetic puzzles, logic problems related to crime detection, and logic and arithmetic puzzles involving King Arthur and his Dogs of the Round Table. 160pp. 5 3/8 x 8 1/2. 0-486-47435-6

UNDECIDABLE THEORIES: Studies in Logic and the Foundation of Mathematics, Alfred Tarski in collaboration with Andrzej Mostowski and Raphael M. Robinson. This well-known book by the famed logician consists of three treatises: "A General Method in Proofs of Undecidability," "Undecidability and Essential Undecidability in Mathematics," and "Undecidability of the Elementary Theory of Groups." 1953 edition. 112pp. 5 3/8 x 8 1/2. 0-486-47703-7

LOGIC FOR MATHEMATICIANS, J. Barkley Rosser. Examination of essential topics and theorems assumes no background in logic. "Undoubtedly a major addition to the literature of mathematical logic." – *Bulletin of the American Mathematical Society.* 1978 edition. 592pp. 6 1/8 x 9 1/4. 0-486-46898-4

INTRODUCTION TO PROOF IN ABSTRACT MATHEMATICS, Andrew Wohlgemuth. This undergraduate text teaches students what constitutes an acceptable proof, and it develops their ability to do proofs of routine problems as well as those requiring creative insights. 1990 edition. 384pp. 6 1/2 x 9 1/4. 0-486-47854-8

FIRST COURSE IN MATHEMATICAL LOGIC, Patrick Suppes and Shirley Hill. Rigorous introduction is simple enough in presentation and context for wide range of students. Symbolizing sentences; logical inference; truth and validity; truth tables; terms, predicates, universal quantifiers; universal specification and laws of identity; more. 288pp. 5 3/8 x 8 1/2. 0-486-42259-3

Browse over 9,000 books at www.doverpublications.com

Mathematics–Algebra and Calculus

VECTOR CALCULUS, Peter Baxandall and Hans Liebeck. This introductory text offers a rigorous, comprehensive treatment. Classical theorems of vector calculus are amply illustrated with figures, worked examples, physical applications, and exercises with hints and answers. 1986 edition. 560pp. 5 3/8 x 8 1/2. 0-486-46620-5

ADVANCED CALCULUS: An Introduction to Classical Analysis, Louis Brand. A course in analysis that focuses on the functions of a real variable, this text introduces the basic concepts in their simplest setting and illustrates its teachings with numerous examples, theorems, and proofs. 1955 edition. 592pp. 5 3/8 x 8 1/2. 0-486-44548-8

ADVANCED CALCULUS, Avner Friedman. Intended for students who have already completed a one-year course in elementary calculus, this two-part treatment advances from functions of one variable to those of several variables. Solutions. 1971 edition. 432pp. 5 3/8 x 8 1/2. 0-486-45795-8

METHODS OF MATHEMATICS APPLIED TO CALCULUS, PROBABILITY, AND STATISTICS, Richard W. Hamming. This 4-part treatment begins with algebra and analytic geometry and proceeds to an exploration of the calculus of algebraic functions and transcendental functions and applications. 1985 edition. Includes 310 figures and 18 tables. 880pp. 6 1/2 x 9 1/4. 0-486-43945-3

BASIC ALGEBRA I: Second Edition, Nathan Jacobson. A classic text and standard reference for a generation, this volume covers all undergraduate algebra topics, including groups, rings, modules, Galois theory, polynomials, linear algebra, and associative algebra. 1985 edition. 528pp. 6 1/8 x 9 1/4. 0-486-47189-6

BASIC ALGEBRA II: Second Edition, Nathan Jacobson. This classic text and standard reference comprises all subjects of a first-year graduate-level course, including in-depth coverage of groups and polynomials and extensive use of categories and functors. 1989 edition. 704pp. 6 1/8 x 9 1/4. 0-486-47187-X

CALCULUS: An Intuitive and Physical Approach (Second Edition), Morris Kline. Application-oriented introduction relates the subject as closely as possible to science with explorations of the derivative; differentiation and integration of the powers of x; theorems on differentiation, antidifferentiation; the chain rule; trigonometric functions; more. Examples. 1967 edition. 960pp. 6 1/2 x 9 1/4. 0-486-40453-6

ABSTRACT ALGEBRA AND SOLUTION BY RADICALS, John E. Maxfield and Margaret W. Maxfield. Accessible advanced undergraduate-level text starts with groups, rings, fields, and polynomials and advances to Galois theory, radicals and roots of unity, and solution by radicals. Numerous examples, illustrations, exercises, appendixes. 1971 edition. 224pp. 6 1/8 x 9 1/4. 0-486-47723-1

AN INTRODUCTION TO THE THEORY OF LINEAR SPACES, Georgi E. Shilov. Translated by Richard A. Silverman. Introductory treatment offers a clear exposition of algebra, geometry, and analysis as parts of an integrated whole rather than separate subjects. Numerous examples illustrate many different fields, and problems include hints or answers. 1961 edition. 320pp. 5 3/8 x 8 1/2. 0-486-63070-6

LINEAR ALGEBRA, Georgi E. Shilov. Covers determinants, linear spaces, systems of linear equations, linear functions of a vector argument, coordinate transformations, the canonical form of the matrix of a linear operator, bilinear and quadratic forms, and more. 387pp. 5 3/8 x 8 1/2. 0-486-63518-X

Mathematics–Probability and Statistics

BASIC PROBABILITY THEORY, Robert B. Ash. This text emphasizes the probabilistic way of thinking, rather than measure-theoretic concepts. Geared toward advanced undergraduates and graduate students, it features solutions to some of the problems. 1970 edition. 352pp. 5 3/8 x 8 1/2. 0-486-46628-0

PRINCIPLES OF STATISTICS, M. G. Bulmer. Concise description of classical statistics, from basic dice probabilities to modern regression analysis. Equal stress on theory and applications. Moderate difficulty; only basic calculus required. Includes problems with answers. 252pp. 5 5/8 x 8 1/4. 0-486-63760-3

OUTLINE OF BASIC STATISTICS: Dictionary and Formulas, John E. Freund and Frank J. Williams. Handy guide includes a 70-page outline of essential statistical formulas covering grouped and ungrouped data, finite populations, probability, and more, plus over 1,000 clear, concise definitions of statistical terms. 1966 edition. 208pp. 5 3/8 x 8 1/2. 0-486-47769-X

GOOD THINKING: The Foundations of Probability and Its Applications, Irving J. Good. This in-depth treatment of probability theory by a famous British statistician explores Keynesian principles and surveys such topics as Bayesian rationality, corroboration, hypothesis testing, and mathematical tools for induction and simplicity. 1983 edition. 352pp. 5 3/8 x 8 1/2. 0-486-47438-0

INTRODUCTION TO PROBABILITY THEORY WITH CONTEMPORARY APPLICATIONS, Lester L. Helms. Extensive discussions and clear examples, written in plain language, expose students to the rules and methods of probability. Exercises foster problem-solving skills, and all problems feature step-by-step solutions. 1997 edition. 368pp. 6 1/2 x 9 1/4. 0-486-47418-6

CHANCE, LUCK, AND STATISTICS, Horace C. Levinson. In simple, non-technical language, this volume explores the fundamentals governing chance and applies them to sports, government, and business. "Clear and lively ... remarkably accurate." – *Scientific Monthly.* 384pp. 5 3/8 x 8 1/2. 0-486-41997-5

FIFTY CHALLENGING PROBLEMS IN PROBABILITY WITH SOLUTIONS, Frederick Mosteller. Remarkable puzzlers, graded in difficulty, illustrate elementary and advanced aspects of probability. These problems were selected for originality, general interest, or because they demonstrate valuable techniques. Also includes detailed solutions. 88pp. 5 3/8 x 8 1/2. 0-486-65355-2

EXPERIMENTAL STATISTICS, Mary Gibbons Natrella. A handbook for those seeking engineering information and quantitative data for designing, developing, constructing, and testing equipment. Covers the planning of experiments, the analyzing of extreme-value data; and more. 1966 edition. Index. Includes 52 figures and 76 tables. 560pp. 8 3/8 x 11. 0-486-43937-2

STOCHASTIC MODELING: Analysis and Simulation, Barry L. Nelson. Coherent introduction to techniques also offers a guide to the mathematical, numerical, and simulation tools of systems analysis. Includes formulation of models, analysis, and interpretation of results. 1995 edition. 336pp. 6 1/8 x 9 1/4. 0-486-47770-3

INTRODUCTION TO BIOSTATISTICS: Second Edition, Robert R. Sokal and F. James Rohlf. Suitable for undergraduates with a minimal background in mathematics, this introduction ranges from descriptive statistics to fundamental distributions and the testing of hypotheses. Includes numerous worked-out problems and examples. 1987 edition. 384pp. 6 1/8 x 9 1/4. 0-486-46961-1

Mathematics–Geometry and Topology

PROBLEMS AND SOLUTIONS IN EUCLIDEAN GEOMETRY, M. N. Aref and William Wernick. Based on classical principles, this book is intended for a second course in Euclidean geometry and can be used as a refresher. More than 200 problems include hints and solutions. 1968 edition. 272pp. 5 3/8 x 8 1/2. 0-486-47720-7

TOPOLOGY OF 3-MANIFOLDS AND RELATED TOPICS, Edited by M. K. Fort, Jr. With a New Introduction by Daniel Silver. Summaries and full reports from a 1961 conference discuss decompositions and subsets of 3-space; n-manifolds; knot theory; the Poincaré conjecture; and periodic maps and isotopies. Familiarity with algebraic topology required. 1962 edition. 272pp. 6 1/8 x 9 1/4. 0-486-47753-3

POINT SET TOPOLOGY, Steven A. Gaal. Suitable for a complete course in topology, this text also functions as a self-contained treatment for independent study. Additional enrichment materials make it equally valuable as a reference. 1964 edition. 336pp. 5 3/8 x 8 1/2. 0-486-47222-1

INVITATION TO GEOMETRY, Z. A. Melzak. Intended for students of many different backgrounds with only a modest knowledge of mathematics, this text features self-contained chapters that can be adapted to several types of geometry courses. 1983 edition. 240pp. 5 3/8 x 8 1/2. 0-486-46626-4

TOPOLOGY AND GEOMETRY FOR PHYSICISTS, Charles Nash and Siddhartha Sen. Written by physicists for physics students, this text assumes no detailed background in topology or geometry. Topics include differential forms, homotopy, homology, cohomology, fiber bundles, connection and covariant derivatives, and Morse theory. 1983 edition. 320pp. 5 3/8 x 8 1/2. 0-486-47852-1

BEYOND GEOMETRY: Classic Papers from Riemann to Einstein, Edited with an Introduction and Notes by Peter Pesic. This is the only English-language collection of these 8 accessible essays. They trace seminal ideas about the foundations of geometry that led to Einstein's general theory of relativity. 224pp. 6 1/8 x 9 1/4. 0-486-45350-2

GEOMETRY FROM EUCLID TO KNOTS, Saul Stahl. This text provides a historical perspective on plane geometry and covers non-neutral Euclidean geometry, circles and regular polygons, projective geometry, symmetries, inversions, informal topology, and more. Includes 1,000 practice problems. Solutions available. 2003 edition. 480pp. 6 1/8 x 9 1/4. 0-486-47459-3

TOPOLOGICAL VECTOR SPACES, DISTRIBUTIONS AND KERNELS, François Trèves. Extending beyond the boundaries of Hilbert and Banach space theory, this text focuses on key aspects of functional analysis, particularly in regard to solving partial differential equations. 1967 edition. 592pp. 5 3/8 x 8 1/2. 0-486-45352-9

INTRODUCTION TO PROJECTIVE GEOMETRY, C. R. Wylie, Jr. This introductory volume offers strong reinforcement for its teachings, with detailed examples and numerous theorems, proofs, and exercises, plus complete answers to all odd-numbered end-of-chapter problems. 1970 edition. 576pp. 6 1/8 x 9 1/4. 0-486-46895-X

FOUNDATIONS OF GEOMETRY, C. R. Wylie, Jr. Geared toward students preparing to teach high school mathematics, this text explores the principles of Euclidean and non-Euclidean geometry and covers both generalities and specifics of the axiomatic method. 1964 edition. 352pp. 6 x 9. 0-486-47214-0

Mathematics-History

THE WORKS OF ARCHIMEDES, Archimedes. Translated by Sir Thomas Heath. Complete works of ancient geometer feature such topics as the famous problems of the ratio of the areas of a cylinder and an inscribed sphere; the properties of conoids, spheroids, and spirals; more. 326pp. 5 3/8 x 8 1/2. 0-486-42084-1

THE HISTORICAL ROOTS OF ELEMENTARY MATHEMATICS, Lucas N. H. Bunt, Phillip S. Jones, and Jack D. Bedient. Exciting, hands-on approach to understanding fundamental underpinnings of modern arithmetic, algebra, geometry and number systems examines their origins in early Egyptian, Babylonian, and Greek sources. 336pp. 5 3/8 x 8 1/2. 0-486-25563-8

THE THIRTEEN BOOKS OF EUCLID'S ELEMENTS, Euclid. Contains complete English text of all 13 books of the Elements plus critical apparatus analyzing each definition, postulate, and proposition in great detail. Covers textual and linguistic matters; mathematical analyses of Euclid's ideas; classical, medieval, Renaissance and modern commentators; refutations, supports, extrapolations, reinterpretations and historical notes. 995 figures. Total of 1,425pp. All books 5 3/8 x 8 1/2.
Vol. I: 443pp. 0-486-60088-2
Vol. II: 464pp. 0-486-60089-0
Vol. III: 546pp. 0-486-60090-4

A HISTORY OF GREEK MATHEMATICS, Sir Thomas Heath. This authoritative two-volume set that covers the essentials of mathematics and features every landmark innovation and every important figure, including Euclid, Apollonius, and others. 5 3/8 x 8 1/2. Vol. I: 461pp. 0-486-24073-8
Vol. II: 597pp. 0-486-24074-6

A MANUAL OF GREEK MATHEMATICS, Sir Thomas L. Heath. This concise but thorough history encompasses the enduring contributions of the ancient Greek mathematicians whose works form the basis of most modern mathematics. Discusses Pythagorean arithmetic, Plato, Euclid, more. 1931 edition. 576pp. 5 3/8 x 8 1/2.
0-486-43231-9

CHINESE MATHEMATICS IN THE THIRTEENTH CENTURY, Ulrich Libbrecht. An exploration of the 13th-century mathematician Ch'in, this fascinating book combines what is known of the mathematician's life with a history of his only extant work, the Shu-shu chiu-chang. 1973 edition. 592pp. 5 3/8 x 8 1/2.
0-486-44619-0

PHILOSOPHY OF MATHEMATICS AND DEDUCTIVE STRUCTURE IN EUCLID'S ELEMENTS, Ian Mueller. This text provides an understanding of the classical Greek conception of mathematics as expressed in Euclid's Elements. It focuses on philosophical, foundational, and logical questions and features helpful appendixes. 400pp. 6 1/2 x 9 1/4. 0-486-45300-6

BEYOND GEOMETRY: Classic Papers from Riemann to Einstein, Edited with an Introduction and Notes by Peter Pesic. This is the only English-language collection of these 8 accessible essays. They trace seminal ideas about the foundations of geometry that led to Einstein's general theory of relativity. 224pp. 6 1/8 x 9 1/4. 0-486-45350-2

HISTORY OF MATHEMATICS, David E. Smith. Two-volume history – from Egyptian papyri and medieval maps to modern graphs and diagrams. Non-technical chronological survey with thousands of biographical notes, critical evaluations, and contemporary opinions on over 1,100 mathematicians. 5 3/8 x 8 1/2.
Vol. I: 618pp. 0-486-20429-4
Vol. II: 736pp. 0-486-20430-8

Physics

THEORETICAL NUCLEAR PHYSICS, John M. Blatt and Victor F. Weisskopf. An uncommonly clear and cogent investigation and correlation of key aspects of theoretical nuclear physics by leading experts: the nucleus, nuclear forces, nuclear spectroscopy, two-, three- and four-body problems, nuclear reactions, beta-decay and nuclear shell structure. 896pp. 5 3/8 x 8 1/2. 0-486-66827-4

QUANTUM THEORY, David Bohm. This advanced undergraduate-level text presents the quantum theory in terms of qualitative and imaginative concepts, followed by specific applications worked out in mathematical detail. 655pp. 5 3/8 x 8 1/2. 0-486-65969-0

ATOMIC PHYSICS AND HUMAN KNOWLEDGE, Niels Bohr. Articles and speeches by the Nobel Prize–winning physicist, dating from 1934 to 1958, offer philosophical explorations of the relevance of atomic physics to many areas of human endeavor. 1961 edition. 112pp. 5 3/8 x 8 1/2. 0-486-47928-5

COSMOLOGY, Hermann Bondi. A co-developer of the steady-state theory explores his conception of the expanding universe. This historic book was among the first to present cosmology as a separate branch of physics. 1961 edition. 192pp. 5 3/8 x 8 1/2. 0-486-47483-6

LECTURES ON QUANTUM MECHANICS, Paul A. M. Dirac. Four concise, brilliant lectures on mathematical methods in quantum mechanics from Nobel Prize-winning quantum pioneer build on idea of visualizing quantum theory through the use of classical mechanics. 96pp. 5 3/8 x 8 1/2. 0-486-41713-1

THE PRINCIPLE OF RELATIVITY, Albert Einstein and Frances A. Davis. Eleven papers that forged the general and special theories of relativity include seven papers by Einstein, two by Lorentz, and one each by Minkowski and Weyl. 1923 edition. 240pp. 5 3/8 x 8 1/2. 0-486-60081-5

PHYSICS OF WAVES, William C. Elmore and Mark A. Heald. Ideal as a classroom text or for individual study, this unique one-volume overview of classical wave theory covers wave phenomena of acoustics, optics, electromagnetic radiations, and more. 477pp. 5 3/8 x 8 1/2. 0-486-64926-1

THERMODYNAMICS, Enrico Fermi. In this classic of modern science, the Nobel Laureate presents a clear treatment of systems, the First and Second Laws of Thermodynamics, entropy, thermodynamic potentials, and much more. Calculus required. 160pp. 5 3/8 x 8 1/2. 0-486-60361-X

QUANTUM THEORY OF MANY-PARTICLE SYSTEMS, Alexander L. Fetter and John Dirk Walecka. Self-contained treatment of nonrelativistic many-particle systems discusses both formalism and applications in terms of ground-state (zero-temperature) formalism, finite-temperature formalism, canonical transformations, and applications to physical systems. 1971 edition. 640pp. 5 3/8 x 8 1/2. 0-486-42827-3

QUANTUM MECHANICS AND PATH INTEGRALS: Emended Edition, Richard P. Feynman and Albert R. Hibbs. Emended by Daniel F. Styer. The Nobel Prize–winning physicist presents unique insights into his theory and its applications. Feynman starts with fundamentals and advances to the perturbation method, quantum electrodynamics, and statistical mechanics. 1965 edition, emended in 2005. 384pp. 6 1/8 x 9 1/4. 0-486-47722-3

Physics

INTRODUCTION TO MODERN OPTICS, Grant R. Fowles. A complete basic undergraduate course in modern optics for students in physics, technology, and engineering. The first half deals with classical physical optics; the second, quantum nature of light. Solutions. 336pp. 5 3/8 x 8 1/2. 0-486-65957-7

THE QUANTUM THEORY OF RADIATION: Third Edition, W. Heitler. The first comprehensive treatment of quantum physics in any language, this classic introduction to basic theory remains highly recommended and widely used, both as a text and as a reference. 1954 edition. 464pp. 5 3/8 x 8 1/2. 0-486-64558-4

QUANTUM FIELD THEORY, Claude Itzykson and Jean-Bernard Zuber. This comprehensive text begins with the standard quantization of electrodynamics and perturbative renormalization, advancing to functional methods, relativistic bound states, broken symmetries, nonabelian gauge fields, and asymptotic behavior. 1980 edition. 752pp. 6 1/2 x 9 1/4. 0-486-44568-2

FOUNDATIONS OF POTENTIAL THERY, Oliver D. Kellogg. Introduction to fundamentals of potential functions covers the force of gravity, fields of force, potentials, harmonic functions, electric images and Green's function, sequences of harmonic functions, fundamental existence theorems, and much more. 400pp. 5 3/8 x 8 1/2. 0-486-60144-7

FUNDAMENTALS OF MATHEMATICAL PHYSICS, Edgar A. Kraut. Indispensable for students of modern physics, this text provides the necessary background in mathematics to study the concepts of electromagnetic theory and quantum mechanics. 1967 edition. 480pp. 6 1/2 x 9 1/4. 0-486-45809-1

GEOMETRY AND LIGHT: The Science of Invisibility, Ulf Leonhardt and Thomas Philbin. Suitable for advanced undergraduate and graduate students of engineering, physics, and mathematics and scientific researchers of all types, this is the first authoritative text on invisibility and the science behind it. More than 100 full-color illustrations, plus exercises with solutions. 2010 edition. 288pp. 7 x 9 1/4. 0-486-47693-6

QUANTUM MECHANICS: New Approaches to Selected Topics, Harry J. Lipkin. Acclaimed as "excellent" (*Nature*) and "very original and refreshing" (*Physics Today*), these studies examine the Mössbauer effect, many-body quantum mechanics, scattering theory, Feynman diagrams, and relativistic quantum mechanics. 1973 edition. 480pp. 5 3/8 x 8 1/2. 0-486-45893-8

THEORY OF HEAT, James Clerk Maxwell. This classic sets forth the fundamentals of thermodynamics and kinetic theory simply enough to be understood by beginners, yet with enough subtlety to appeal to more advanced readers, too. 352pp. 5 3/8 x 8 1/2. 0-486-41735-2

QUANTUM MECHANICS, Albert Messiah. Subjects include formalism and its interpretation, analysis of simple systems, symmetries and invariance, methods of approximation, elements of relativistic quantum mechanics, much more. "Strongly recommended." – *American Journal of Physics*. 1152pp. 5 3/8 x 8 1/2. 0-486-40924-4

RELATIVISTIC QUANTUM FIELDS, Charles Nash. This graduate-level text contains techniques for performing calculations in quantum field theory. It focuses chiefly on the dimensional method and the renormalization group methods. Additional topics include functional integration and differentiation. 1978 edition. 240pp. 5 3/8 x 8 1/2. 0-486-47752-5

Physics

MATHEMATICAL TOOLS FOR PHYSICS, James Nearing. Encouraging students' development of intuition, this original work begins with a review of basic mathematics and advances to infinite series, complex algebra, differential equations, Fourier series, and more. 2010 edition. 496pp. 6 1/8 x 9 1/4. 0-486-48212-X

TREATISE ON THERMODYNAMICS, Max Planck. Great classic, still one of the best introductions to thermodynamics. Fundamentals, first and second principles of thermodynamics, applications to special states of equilibrium, more. Numerous worked examples. 1917 edition. 297pp. 5 3/8 x 8. 0-486-66371-X

AN INTRODUCTION TO RELATIVISTIC QUANTUM FIELD THEORY, Silvan S. Schweber. Complete, systematic, and self-contained, this text introduces modern quantum field theory. "Combines thorough knowledge with a high degree of didactic ability and a delightful style." – *Mathematical Reviews.* 1961 edition. 928pp. 5 3/8 x 8 1/2. 0-486-44228-4

THE ELECTROMAGNETIC FIELD, Albert Shadowitz. Comprehensive undergraduate text covers basics of electric and magnetic fields, building up to electromagnetic theory. Related topics include relativity theory. Over 900 problems, some with solutions. 1975 edition. 768pp. 5 5/8 x 8 1/4. 0-486-65660-8

THE PRINCIPLES OF STATISTICAL MECHANICS, Richard C. Tolman. Definitive treatise offers a concise exposition of classical statistical mechanics and a thorough elucidation of quantum statistical mechanics, plus applications of statistical mechanics to thermodynamic behavior. 1930 edition. 704pp. 5 5/8 x 8 1/4. 0-486-63896-0

INTRODUCTION TO THE PHYSICS OF FLUIDS AND SOLIDS, James S. Trefil. This interesting, informative survey by a well-known science author ranges from classical physics and geophysical topics, from the rings of Saturn and the rotation of the galaxy to underground nuclear tests. 1975 edition. 320pp. 5 3/8 x 8 1/2. 0-486-47437-2

STATISTICAL PHYSICS, Gregory H. Wannier. Classic text combines thermodynamics, statistical mechanics, and kinetic theory in one unified presentation. Topics include equilibrium statistics of special systems, kinetic theory, transport coefficients, and fluctuations. Problems with solutions. 1966 edition. 532pp. 5 3/8 x 8 1/2. 0-486-65401-X

SPACE, TIME, MATTER, Hermann Weyl. Excellent introduction probes deeply into Euclidean space, Riemann's space, Einstein's general relativity, gravitational waves and energy, and laws of conservation. "A classic of physics." – *British Journal for Philosophy and Science.* 330pp. 5 3/8 x 8 1/2. 0-486-60267-2

RANDOM VIBRATIONS: Theory and Practice, Paul H. Wirsching, Thomas L. Paez and Keith Ortiz. Comprehensive text and reference covers topics in probability, statistics, and random processes, plus methods for analyzing and controlling random vibrations. Suitable for graduate students and mechanical, structural, and aerospace engineers. 1995 edition. 464pp. 5 3/8 x 8 1/2. 0-486-45015-5

PHYSICS OF SHOCK WAVES AND HIGH-TEMPERATURE HYDRO DYNAMIC PHENOMENA, Ya B. Zel'dovich and Yu P. Raizer. Physical, chemical processes in gases at high temperatures are focus of outstanding text, which combines material from gas dynamics, shock-wave theory, thermodynamics and statistical physics, other fields. 284 illustrations. 1966–1967 edition. 944pp. 6 1/8 x 9 1/4. 0-486-42002-7